Numerical Linear Algebra

Numerical Linear Algebra

Proceedings of the Conference in Numerical Linear Algebra
and Scientific Computation, Kent (Ohio), USA
March 13–14, 1992

Editors

Lothar Reichel
Arden Ruttan
Richard S. Varga

Walter de Gruyter · Berlin · New York 1993

Editors

Lothar Reichel, Arden Ruttan, Richard S. Varga
Department of Mathematics and Computer Science
Kent State University, Kent, Ohio 44242, USA

1991 *Mathematics Subject Classification: Primary:* 65-06
Secondary: 15A48, 43A10, 65F05, 65F10, 65F15, 65F25, 65F50, 65N20, 65N22, 65Y05
Keywords: Scientific Computation, Iterative Methods, Linear Systems of Equations, Eigenvalue Problems, Parallel Computation, Signal Processing, Image Analysis

⊗ Printed on acid-free paper which falls within the guidelines of the ANSI to ensure permanence and durability.

Library of Congress Cataloging-in-Publication Data

Conference in Numerical Linear Algebra and Scientific Computation (1992 ; Kent, Ohio)
Numerical linear algebra : proceedings of the Conference in Numerical Linear Algebra and Scientific Computation, Kent (Ohio), March 13–14, 1992 / editors, Lothar Reichel, Arden Ruttan, Richard S. Varga.
p. cm.
Includes bibliographical references.
ISBN 3-11-013784-4 (acid-free)
1. Algebras, Linear – Congresses. 2. Numerical analysis – Congresses. I. Reichel, Lothar, 1952– . II. Ruttan, Arden. III. Varga, Richard S. IV. Title.
QA184.C655 1992
512′.5–dc20 93-8437
CIP

Die Deutsche Bibliothek – Cataloging-in-Publication Data

Numerical linear algebra : proceedings of the Conference in Numerical Linear Algebra and Scientific Computation Kent (Ohio), March 13–14, 1992 / ed. Lothar Reichel ... – Berlin ; New York : de Gruyter, 1993
ISBN 3-11-013784-4
NE: Reichel, Lothar [Hrsg.]; Conference in Numerical Linear Algebra and Scientific Computation <1992, Kent, Ohio>

Printed in Germany. Printing: Werner Hildebrand, Berlin. Binding: Lüderitz & Bauer, Berlin.
Cover design: Thomas Bonnie, Hamburg.

PREFACE

On Friday and Saturday, March 13-14, 1992, a meeting of the Western Pennsylvania and Eastern Ohio section of SIAM, with the title Numerical Linear Algebra and Scientific Computing was held at Kent State University. There were approximately 60 participants present, representing the states Illinois, Minnesota, Maryland, Ohio, Pennsylvania, Michigan, New Jersey, Kentucky, and California and the country of Germany.

In the short period of two days, the participants heard new research results in Numerical Linear Algebra from acknowledged leaders in the field. The talks displayed the present activity in this area of Numerical Analysis and illustrated the wide diversity of the ongoing research. Current work in Numerical Linear Algebra spans from theoretical investigations of properties of matrices with a structure of interest in applications, exemplified by the contributions by Nabben and Varga, to the implementation and timing of algorithms on parallel computers, in these Proceedings represented by the paper by Calvetti et al.

Several of the contributions describe and analyze new iterative methods for the solution of large systems of linear equations. The paper by Freund discusses the application of the recently developed powerful quasi-minimal residual (QMR) method to the iterative solution of several linear systems of equations, with the same right-hand sides and in which the matrices only differ by a multiple of the identity. Such linear systems arise, e.g., in problems in control theory, as well as in the solution of time-dependent partial differential equations when the discretization is carried out by schemes that recently have been proposed by Gallopoulos and Saad. The latter schemes are suitable for implementation on parallel computers. The paper by Freund and Nachtigal discusses a new implementation of the QMR method that is based on coupling the recurrence relations, for the computation of a basis of the Krylov subspace, with the computation of the residual vector. This implementation is more stable the the one previously used. It is well known that iterative methods for the solution of large systems of linear equations typically require preconditioning in order to be competitive with direct solution methods. The choice of preconditioner is generally not easy and the determination of new preconditioners continues to be an active area of research. The contribution by Ma and Saad describes the novel use of a block-ADI method for preconditioning linear systems of equations that arise from the discretization of certain partial differential equations. The SOR method of Frankel and Young in the 50's continues to recieve considerable attention by numerical analysts and engineers because of its versatility. The paper by Eiermann and Varga analyzes possibilities of coupling the SOR method with semi-iterative schemes and shows that powerful iterative schemes can be obtained in this manner. One of the most popular iterative schemes for the solution of large systems of linear equations is the GMRES method by Saad and Schultz. The implementation by Saad and Schultz is based on the Arnoldi process and uses the modified Gram-Schmidt method for computing orthogonal bases of certain Krylov subspaces. It can be difficult to implement the modified Gram-Schmidt method efficiently on parallel computers due to its low granularity.

Calvetti et al. describe how the computations of the GMRES method can be organized so as to make efficient parallel computation possible. Illustrative timings on an IBM 3090-600S VF computer are presented. The contribution by Hanke et al. discusses the iterative solution of large linear systems of equations that arise in tomography. These systems are typically very ill-conditioned and this makes the development of iterative methods for such problems highly challenging. Hanke et al. present a new very promising approach to the iterative solution of large ill-conditioned systems of linear equations.

Two of the contributions of these Proceedings are concerned with the computation of eigenvalues and eigenvectors of matrices. The advent of parallel computers has generated interest in divide-and-conquer algorithms for this purpose. The attraction of these algorithms stems from that the computation can be divided into subtasks that can be solved independently and therefore in parallel. Divide-and-conquer algorithms and their parallel implementation have previously been presented for the symmetric eigenvalue problem and the unitary eigenvalue problem. The paper by Carlos and Gragg describes a divide-and-conquer method for a symmetric generalized eigenvalue problem. The computations are carried out by transforming the original problem into an eigenvalue problem for an arrowhead matrix. A new root-finder is presented that yields rapid convergence and high accuracy. The approach of this paper is likely to be important for the further development of divide-and-conquer algorithms.

The qd algorithm by Rutishauser is the concern of the paper by Fernando and Parlett. They organize the computation in a new manner, which they show gives improved accuracy. Their paper illustrates that the qd algorithm can be superior to other available methods for eigenvalue computation.

Real-time signal processing is a rich source of challenging problems in Numerical Linear Algebra. It is important to be able to carry out the computations quickly and to achieve accurate results. This favors the development of parallel algorithms as well as methods that reduce the arithmetic operations necessary by utilizing the structure of the computational problem. Algorithms for computing the singular value decomposition are central in signal processing. The paper by Bojanczyk and Van Dooren describes a new approach to computing the singular value decomposition of a product of matrices, without explicitly forming the product.

It is well known that numerical discretization of partial differential equations of Poisson-type gives rise to coefficient matrices which possess an abundance of nice properties. The matrices obtained are known as Stieltjes matrices and M-matrices. In these Proceedings Nabben gives a generalization of M-matrices, with application to the numerical solution of Euler equations, and Varga and Nabben study extensions of the newly coined symmetric ultrametric matrices, whose inverses turn out to be diagonally dominant Stieltjes matrices.

We are happy to have secured from some of the speakers a number of polished manuscripts on the leading edge of research in numerical analysis. The manuscripts for the lectures are contained in these Proceedings for our conference, and we hope that the readers of these Proceedings will enjoy the content as much as we have.

Lothar Reichel Arden Ruttan Richard S. Varga

CONTENTS

On propagating orthogonal transformations in a product of 2×2 triangular matrices

Adam Bojanczyk[*†] *and Paul Van Dooren*[‡§]

Abstract. In this note, we propose an implicit method for applying orthogonal transformations on both sides of a product of upper triangular 2×2 matrices that preserve upper triangularity of the factors. Such problems arise in Jacobi type methods for computing the PSVD of a product of several matrices, and in ordering eigenvalues in the periodic Schur decomposition.

Key words. orthogonal transformations, SVD, PSVD, Schur decompostion.

AMS(MOS) subject classifications. 15A23, 65F25.

1 Introduction

The problem of computing the singular value decomposition (SVD) of a product of matrices (PSVD) has been considered in [1],[2], [3], [10]. The computation proceeds in two stages. In the first stage the matrices are transformed into the upper triangular forms. In the second iterative stage an implicit Jacobi-type method is applied to the triangular matrices. It is important that after each iteration the matrices stay triangular [8].

A crucial aspect in such implicit Jacobi iterations is the accurate computation of the PSVD of a product of 2×2 triangular matrices. There two conditions have to be satisfied. First, one has to ensure that the orthogonal transformations applied to the triangular matrices must leave the matrices triangular, and second, that the transformations diagonalize the product accurately. It was shown in [1] and

---

[*] Cornell University, Department of Electrical Engineering, Ithaca, NY 14853-3801.

[†] Research supported in part by the Joint Services Electronics Program (Grant F49620–90–C–0039 monitored by AFOSR).

[‡] University of Illinois at Urbana-Champaign, Coordinated Science Lab, Urbana, IL 61801.

[§] Research supported in part by the Research Board of the University of Illinois at Urbana-Champaign (Grant P 1–2–68114) and by the National Science Foundation (Grant CCR 9209349).

Numerical Linear Algebra

[2] that these two conditions are satisfied by a so-called *half-recursive* and *direct* method, respectively, for computing the SVD of the product of two matrices.

In this note we analyze an extension of the *half-recursive* method for computing the SVD of the product of many 2×2 triangular matrices. We also show that the extension of the *half-recursive* method can be used for swapping eigenvalues in the periodic Schur decomposition described in [4]. For simplicity we assume real matrices and real eigenvalues, but all results are easily extended to the complex case.

2 Criterion for numerical triangularity

Suppose we are given k, $k > 1$, upper triangular matrices A_i, $i = 1, 2..., k$,

$$A_i = \begin{pmatrix} a_i & b_i \\ 0 & d_i \end{pmatrix} .$$

We denote the product of A_i, $i = 1, 2..., k$, by A,

$$A = A_1 \cdots A_k = \begin{pmatrix} a & b \\ 0 & d \end{pmatrix} .$$

Let the orthogonal matrices Q_1 and Q_{k+1} be such that

$$A' = Q_1 A Q_{k+1}^T = \begin{pmatrix} a' & b' \\ 0 & d' \end{pmatrix} \tag{2.1}$$

is upper triangular. In case we are interested in finding the *Singular Value Decomposition* of A, one imposes the additional condition that $b' = 0$. This defines uniquely the above decomposition up to permutations that interchange the diagonal elements of A'. In case we are interested in finding the *Schur Form* of A, one imposes the additional condition that $Q_1 = Q_{k+1}$. Again, this defines uniquely the above decomposition up to the ordering of the diagonal elements of A'. In both cases the transformations Q_1 and Q_{k+1} are thus defined by the choice of ordering of diagonal elements in the resulting matrix A'. Our objective now is to find orthogonal matrices Q_j, $j = 2, 3, ..., k$, such that

$$A_i' = Q_i A_i Q_{i+1}^T = \begin{pmatrix} a_i' & b_i' \\ 0 & d_i' \end{pmatrix} \tag{2.2}$$

are meanwhile maintained in upper triangular form as well. It is easy to see that if $abd \neq 0$ then for a given pair of orthogonal transformations Q_1 and Q_{k+1} there exist unique (up to the sign) orthogonal transformations $Q_2,...,Q_k$ such that (2.2) is satisfied. There are many mathematically equivalent strategies of determining $Q_2,...,Q_k$. However, as it was shown in [1], [2] and [3], some strategies may produce numerically significantly different results than other strategies. We will consider a particular method numerically acceptable if the triangular matrices after transformations have been applied to them stay numerically triangular in the sense described below.

Let $\bar{A}$ be the computed A, and let $\bar{Q}_i$, $i = 1, 2, ..., k+1$ be the computed transformations. Define

$$\tilde{A}' := \bar{Q}_1 \bar{A} \bar{Q}_{k+1}^T = \begin{pmatrix} \tilde{a}' & \tilde{b}' \\ \tilde{e}' & \tilde{d}' \end{pmatrix} \tag{2.3}$$

and

$$\tilde{A}_i' := \bar{Q}_i A_i \bar{Q}_{i+1}^T = \begin{pmatrix} \tilde{a}_i' & \tilde{b}_i' \\ \tilde{e}_i' & \tilde{d}_i' \end{pmatrix} . \tag{2.4}$$

Let ϵ denote the relative machine precision. Assume that we are given $\bar{Q}_1$ and $\bar{Q}_{k+1}$ such that

$$|\tilde{e}'| = O(\epsilon||\bar{A}||) \tag{2.5a}$$

We will say that $\tilde{A}_i'$ is numerically triangular if

$$|\tilde{e}_i'| = O(\epsilon||A_i||) , \tag{2.5b}$$

We will propose a method for computing nearly orthogonal $\bar{Q}_i$, $i = 2, ..., k$, for which, under a slightly stronger version of the assumption (2.5a), the (2,1) element $\tilde{e}_i{}'$ of $\tilde{A}_i'$ will satisfy (2.5b). Condition (2.5b) justifies truncating the (2,1) element e_i' of $\tilde{A}_i'$ to zero. Thus, $\tilde{e}'$ is also forced to zero.

3 The Algorithm

Our algorithm is a generalization of the algorithms presented in [1] and [3] for computing the PSVD of two and three matrices respectively. There the orthogonal transformations all had the form

$$Q = \begin{pmatrix} s & c \\ -c & s \end{pmatrix} , \tag{3.1}$$

where $c^2 + s^2 = 1$. As we will build on the results presented in those papers we retain this particular choice of orthogonal transformations. While each transformation Q_i is defined by the cosine-sine pair $c_i = \cos\theta_i$ and $s_i = \sin\theta_i$, we also associate Q_i with the tangent

$$t_i = \tan\theta_i .$$

Given t_i, we can easily recover c_i and s_i using the relations

$$c_i = \frac{1}{\sqrt{1+t_i^2}} \quad \text{and} \quad s_i = t_i c_i . \tag{3.2}$$

Following the exposition in [1], [3], we consider the result of applying the left and right transformations Q_l (for the outer left transformation) and Q_r (for the outer right transformation) to a 2×2 upper triangular matrix A:

$$A' = Q_l A Q_r^T = \begin{pmatrix} a' & b' \\ e' & d' \end{pmatrix} = \begin{pmatrix} s_l & c_l \\ -c_l & s_l \end{pmatrix} \begin{pmatrix} a & b \\ 0 & d \end{pmatrix} \begin{pmatrix} s_r & c_r \\ -c_r & s_r \end{pmatrix}^T . \tag{3.3}$$

We can derive from (3.3) these four relations:

$$e' = c_l c_r(-at_r + dt_l - b) , \tag{3.4a}$$

$$b' = c_l c_r(-at_l + dt_r + bt_l t_r) , \tag{3.4b}$$

$$a' = c_l c_r(bt_l + d + at_l t_r) , \tag{3.4c}$$

$$d' = c_l c_r(a - bt_r + dt_l t_r) , \tag{3.4d}$$

where $t_l = \tan\theta_l$ and $t_r = \tan\theta_r$.

The postulates that both e' and b' be zeros define two conditions on t_l and t_r, so that (3.3) represents an SVD of A [5]. The postulate that e' be zero and $t_l = t_r$ represent conditions for swapping eigenvalues of A.

The postulate that e' be zero defines a condition relating θ_l to θ_r, so that if one is known the other can be computed in order to reduce A' to an upper triangular form. For ease of exposition, we assume for now on that $abd \neq 0$. It implies that $c_l c_r \neq 0$, and so the postulate that $e' = 0$ in (3.4a) becomes

$$-at_r + dt_l - b = 0 . \tag{3.5}$$

The consequence of (3.5) is that (3.4c) and (3.4d) simplify to

$$a' = c_l c_r(t_l^2 + 1)d \tag{3.6a}$$

and

$$d' = c_l c_r(t_r^2 + 1)a , \tag{3.6b}$$

respectively.

Assume that $Q_l = Q_1$ and $Q_r = Q_{k+1}$ are given, that is $t_l = t_1$ and $t_r = t_{k+1}$ are known. We will use relations of the type (3.5) with t_l and t_r as the reference tangents to compute the remaining transformations.

Our algorithm can be described recursively as follows. We split the sequence A_1, A_2,...,A_{k+1} into two subsequences of consecutive matrices A_1, A_2,...,A_m and A_{m+1}, A_{m+2},...,A_{k+1} where $1 < m < k+1$. Let us denote

$$A_l \equiv \begin{pmatrix} a_l & b_l \\ 0 & d_l \end{pmatrix} = \prod_{i=1}^{m} A_i \text{ and } A_r \equiv \begin{pmatrix} a_r & b_r \\ 0 & d_r \end{pmatrix} = \prod_{i=m+1}^{k} A_{i+1} . \tag{3.7}$$

Suppose that

$$|t_l d| \leq |t_r a| .$$

Then we propose to compute t_m from the condition (3.5) by the forward substitution,

$$t_m = \frac{d_l t_l - b_l}{a_l} \tag{3.8a}$$

Otherwise, that is when

$$|t_l d| > |t_r a| ,$$

we propose to compute t_m from (3.5) by the backward substitution,

$$t_m = \frac{a_r t_r + b_r}{d_r} . \tag{3.8b}$$

Having defined the first step, the procedure can now be applied recursively to generate all the remaining orthogonal transformations Q_i, $i = 2, ..., k$. Note that there is a lot of freedom in splitting the sequence A_1, A_2,...,A_{k+1} into subsequent subsequences. This might be advantageous for a divide-and-conquer type of computation in a parallel environment.

As will be shown later, under mild conditions on Q_1 and Q_{k+1}, this particular way of generating orthogonal transformations Q_i, $i = 2, ..., k$, will guarantee that all A_i' will be numerically upper triangular in the sense that (2.5b) will be satisfied.

4 Error Analysis

In our error analysis, we adopt a convention that involves a liberal use of Greek letters. For example, by α we mean a relative perturbation of an absolute magnitude not greater than ϵ, where ϵ denotes the machine precision. All terms of order ϵ^2 or higher will be ignored in this first-order analysis.

The function $\mathrm{fl}(a)$ will denote the floating point approximation of a. For the purpose of the analysis, a "bar" denotes a computed quantity which is perturbed as the result of inexact arithmetic. For example, instead of a, b and d, we have the perturbed values $\bar{a}$, $\bar{b}$ and $\bar{d}$ which result from floating point computation of $\prod_{i=1}^{k+1} A_i$. We assume that exact arithmetic may be performed using these perturbed values. The "tilde" symbol is used to denote conceptual values computed exactly from perturbed data.

We start our procedure by computing elements of the product matrix A as the product of $\bar{A}_l$ and $\bar{A}_r$ defined by (3.7):

$$\bar{a} := \mathrm{fl}(\bar{a}_l\bar{a}_r) = \bar{a}_l\bar{a}_r(1+\alpha) , \tag{4.1a}$$

$$\bar{d} := \mathrm{fl}(\bar{d}_l\bar{d}_r) = \bar{d}_l\bar{d}_r(1+\delta) , \tag{4.1b}$$

$$\bar{b} := \mathrm{fl}(\bar{a}_l\bar{b}_r + \bar{b}_l\bar{d}_r) = \bar{a}_l\bar{b}_r(1+2\beta_1) + \bar{b}_l\bar{d}_l(1+2\beta_2) , \tag{4.1c}$$

where, according to our convention, the parameters α, δ, β_1, β_2, and β_3 are all quantities whose absolute values are bounded by ϵ.

Now we specify the condition that we impose on the computed Q_1 and Q_{k+1}.

Assumption I. Throught the rest of this note we will assume that the computed tangents $\bar{t}_l$ and $\bar{t}_r$ corresponding to the outer transformations $Q_l = Q_1$ and $Q_r = Q_{k+1}$ satisfy the following equality

$$\bar{a}(1+C\psi)\bar{t}_r - \bar{d}(1+C\phi)\bar{t}_l + \bar{b}(1+C\chi) = 0 , \tag{4.2a}$$

where $C = C(k)$.

□

Lemma 4.1. *The recurrence (3.8a) yields $\bar{t}_m$ such that*

$$\bar{a}_l(1+2\psi_1)\bar{t}_m - \bar{d}_l(1+\phi_1)\bar{t}_l + \bar{b}_l = 0 . \tag{4.3}$$

Likewise, the recurrence (3.8b) yields $\bar{t}_m$ *such that*

$$\bar{d}_r(1+2\phi_2)\bar{t}_m - \bar{a}_r(1+\psi_2)\bar{t}_r - \bar{b}_r = 0 \,. \tag{4.4}$$

Proof. The proof easily follows from (3.8a) and (3.8b).□

Theorem4.2. *If* $|\bar{t}_l\bar{d}| < |\bar{t}_r\bar{a}|$ *and if* $\bar{t}_m$ *is computed via (3.8a) then* $\bar{t}_m$ *satisfies the relation*

$$\bar{a}_r(1+C_l\psi_l)\bar{t}_r - \bar{d}_r(1+C_l\phi_l)\bar{t}_m + \bar{b}_r(1+C_l\psi_l) = 0 \tag{4.5a}$$

where $C_l = C_l(k)$. *Likewise, if* $|\bar{t}_l\bar{d}| \geq |\bar{t}_r\bar{a}|$ *and if* $\bar{t}_m$ *is computed via (3.8b) then* $\bar{t}_m$ *satisfies the relation*

$$\bar{a}_l(1+C_r\psi_r)\bar{t}_m - \bar{d}_l(1+C_r\phi_r)\bar{t}_l + \bar{b}_l(1+C_r\chi_r) = 0 \tag{4.5b}$$

where $C_r = C_r(k)$.

Proof. We give a proof of the relation (4.5a) only as the relation (4.5b) can be proved in an analogous way.

First from (4.3)-(4.4) we get

$$\bar{a}_l(1+2\psi_1)\bar{t}_m - \bar{d}_l(1+\phi_1)\bar{t}_l + \bar{b}_l = 0 \,, \tag{4.6a}$$

while from Assumption I and (4.1a)-(4.1c) we have

$$\begin{aligned}&\bar{a}_l\bar{a}_r(1+\alpha+C\psi)\bar{t}_r - \bar{d}_l\bar{d}_r(1+\delta+C\phi)\bar{t}_l + \\ &\bar{a}_l\bar{b}_r(1+2\beta_1+C\chi) + \bar{b}_l\bar{d}_r(1+2\beta_2+C\chi) = 0 \,.\end{aligned} \tag{4.6b}$$

By multiplying both sides of (4.6a) by $d_r(1+2\beta_2+C\chi)$ and subtracting from (4.6b) we obtain

$$\begin{aligned}\bar{a}_l\{\bar{a}_r(1+\alpha+C\psi)\bar{t}_r - \bar{a}_r\left(\frac{\bar{d}_l\bar{d}_r}{\bar{a}_l\bar{a}_r}\right)(\delta+C\phi-\phi_1-2\beta_2-C\chi)\bar{t}_l + \\ \bar{b}_r(1+2\beta_1+C\chi) - \bar{d}_r(1+2\beta_2+C\chi+2\psi_1)\bar{t}_m\} = 0 \,,\end{aligned}$$

or, since $\bar{a}_l \neq 0$,

$$\begin{aligned}\bar{a}_r(1+\alpha+C\psi)\bar{t}_r - \bar{a}_r\bar{t}_r\left(\frac{\bar{d}\bar{t}_l}{\bar{a}\bar{t}_r}\right)(\delta+C\phi-\phi_1+2\beta_2+C\chi) + \\ \bar{b}_r(1+2\beta_1+C\chi) - \bar{d}_r(1+2\beta_2+2\psi_1+C\chi)\bar{t}_m = 0 \,.\end{aligned}$$

As we assumed that $|\bar{t}_l\bar{d}| < |\bar{t}_r\bar{a}|$, the above can be rewritten as

$$\bar{a}_r(1+C_l\psi_l)\bar{t}_r - \bar{d}_r(1+C_l\phi_l)\bar{t}_m + \bar{b}_r(1+C_l\chi_l) = 0 \tag{4.7}$$

where $C_l = C_l(k)$ completing the proof.□

We now justify why the (2,1) element in the computed matrix A_i' can be set to zero. Let the cosine and sine pairs $\tilde{c}_i$ and $\tilde{s}_i$ satisfy $\bar{t}_i = \tilde{s}_i/\tilde{c}_i$, for $i = l, m, r$. From (4.2) we can derive that

$$\bar{c}_i := \mathrm{fl}(\tilde{c}_i) = \tilde{c}_i(1+3\mu_i) \,, \tag{4.8a}$$

$$\bar{s}_i := \mathrm{fl}(\tilde{s}_i) = \tilde{s}_i(1+4\nu_i) \,. \tag{4.8b}$$

Let $\tilde{A}_i'$ denote the exact updated matrix derived from $\bar{A}_i$, $i = l, r$, and $\bar{c}_i$, $\bar{s}_i$, $i = l, m, r$ that is

$$\tilde{A}_l' = \begin{pmatrix} \bar{s}_l & \bar{c}_l \\ -\bar{c}_l & \bar{s}_l \end{pmatrix} \begin{pmatrix} \bar{a}_l & \bar{b}_l \\ 0 & \bar{d}_l \end{pmatrix} \begin{pmatrix} \bar{s}_m & -\bar{c}_m \\ \bar{c}_m & \bar{s}_m \end{pmatrix} , \tag{4.9a}$$

and

$$\tilde{A}_r' = \begin{pmatrix} \bar{s}_m & \bar{c}_m \\ -\bar{c}_m & \bar{s}_m \end{pmatrix} \begin{pmatrix} \bar{a}_r & \bar{b}_r \\ 0 & \bar{d}_r \end{pmatrix} \begin{pmatrix} \bar{s}_r & -\bar{c}_r \\ \bar{c}_r & \bar{s}_r \end{pmatrix} . \tag{4.9b}$$

Our next result is a direct consequence of Theorem 4.2 and provides bounds on the elements $\tilde{e}_i'$, $i = l, r$, defined by the relations

$$\tilde{e}_l' := -\bar{c}_l \bar{s}_m a_l + \bar{s}_l \bar{c}_m d_l - \bar{c}_l \bar{c}_m b_l , \tag{4.10a}$$

$$\tilde{e}_r' := -\bar{c}_m \bar{s}_r a_r + \bar{s}_m \bar{c}_r d_r - \bar{c}_m \bar{c}_r b_r . \tag{4.10b}$$

Corollary 4.4: *If $|\bar{t}_l \bar{d}| < |\bar{t}_r \bar{a}|$ and if $\bar{t}_m$ is computed via (3.8a) or if $|\bar{t}_l \bar{d}| \geq |\bar{t}_r \bar{a}|$ and if $\bar{t}_m$ is computed via (3.8b) then*

$$|\tilde{e}_i'| \leq K_i \epsilon ||\bar{A}_i|| \;\; , \text{ for } i = l, r . \tag{4.11}$$

Proof. We prove the corollary for the case when $|\bar{t}_l \bar{d}| < |\bar{t}_r \bar{a}|$ and when $\bar{t}_m$ is computed via (3.8a). The other case can be proved in an analogous manner.

Using (4.3a) we can rewrite (4.10a) as

$$\tilde{e}_l' = -\bar{c}_l \bar{s}_m \bar{a}_l + \bar{s}_l \bar{c}_m \bar{d}_l - \bar{c}_l \bar{c}_m \bar{b}_l +$$

$$\bar{c}_l \bar{c}_m (\bar{a}_l (1 + 2\psi_1) \bar{t}_m - \bar{d}_l (1 + \phi_1) \bar{t}_l + \bar{b}_l) \tag{4.12}$$

from which it follows that

$$|\tilde{e}_l'| \leq K_l \epsilon ||\bar{A}_l|| .$$

Similarly, using (4.5a) we can rewrite (4.10b) as

$$\tilde{e}_r' := -\bar{c}_m \bar{s}_r a_r + \bar{s}_m \bar{c}_r d_r - \bar{c}_r \bar{c}_m b_r +$$

$$\bar{c}_r \bar{c}_m (\bar{a}_r (1 + C_l \psi_l) \bar{t}_r - \bar{d}_r (1 + C_l \phi_l) \bar{t}_m + \bar{b}_r (1 + C_l \psi_l)) \tag{4.13}$$

and thus

$$|\tilde{e}_r'| \leq K_r \epsilon ||\bar{A}_r|| ,$$

completing the proof of (4.10a).□

5 Numerical examples

The SVD algorithms for 2×2 upper triangular matrices in [1],[2] or [5] give $\bar{t}_l$ and $\bar{t}_r$ which satisfy Assumption I. We will illustrate that by using our new scheme triangularity of the transformed factors is preserved.

Consider the case of three matrices in the product. Assume that the given data matrices are

$$A_1 = \begin{pmatrix} 2.316797292247488e+00 & -1.437687878748196e-01 \\ 0 & -2.718295063593277e-02 \end{pmatrix},$$

$$A_2 = \begin{pmatrix} 1.222222234444442e+00 & 3.480474357220011e-01 \\ 0 & 5.674165405829751e+00 \end{pmatrix},$$

$$A_3 = \begin{pmatrix} 2.222222211111111e-01 & 1.732050807568877e+00 \\ 0 & 1.111111110000000e-12 \end{pmatrix}.$$

They generate the matrix product $\bar{A} := A1 \cdot A2 \cdot A3$

$$\bar{A} = \begin{pmatrix} 6.292535886949669e-01 & 4.904546363614013e+00 \\ 0 & -1.713783977472744e-13 \end{pmatrix}.$$

We are interested in computing orthogonal transformations Q_1, Q_2, Q_3 and Q_4 which satisfy (2.2) and (2.1) with the (1,2) element zero. The SVD algorithm for the 2×2 upper triangular matrix $\bar{A}$ in [1] or [5] gives $\bar{t}_1 = 3.437688760727056e-14$ and $\bar{t}_4 = -7.794228673031074e+00$ which satisfy Assumption I. In fact we have

$$\bar{Q}_1 \bar{A} \bar{Q}_4 = \begin{pmatrix} -2.180909253067911e-14 & -7.494178599599612e-30 \\ 0 & 4.944748235423613e+00 \end{pmatrix}$$

We split $\bar{A}$ into the product of $A_{1,2} = A_1 A_2$ and A_3. We note that the ratio

$$\frac{\bar{t}_1 \bar{d}}{\bar{t}_4 \bar{a}} = 1.201223412093697e-27$$

If we compute t_3 from t_1 as indicated by the ratio, and next t_2 as specified by (3.8a) or (3.8b), then Corollary 5.4 will guarantee that the transformed factors will stay (numerically) triangular. Suppose however that we compute t_3 from t_4 and next t_2 from t_3. Then Lemma 4.1 will guarantee that $Q_2 A_2 Q_3^T$ and $Q_3 A_3 Q_4^T$ will stay numerically triangular. However, for the computed $Q_1 A_1 Q_2^T$ we have

$$Q_1 A_1 Q_2^T = \begin{pmatrix} -2.713066430028558e-02 & -1.685188387402401e-03 \\ -1.360106941575845e-04 & 2.321253786046106e+00 \end{pmatrix}$$

which cannot be considered upper triangular. An error of order 10^{-4} has to be introduced to truncate the (2,1) element in $Q_1 A_1 Q_2^T$ so it becomes upper triangular.

References

[1] G.E. Adams, A.W. Bojanczyk and F.T. Luk, *Computing the PSVD of Two 2×2 Triangular Matrices*, submitted SIAM J. Matrix Anal. Appl.

[2] Z. Bai and J.W. Demmel, *Computing the Generalized Singular Value Decomposition*, Report No UCB/CSD 91/645, Computer Science Division, University of California, Berkeley, August 1991.

[3] A.W. Bojanczyk, L.M. Ewerbring, F.T. Luk and P. Van Dooren, *An Accurate Product SVD Algorithm, Signal Processing*, 25 (1991), pp. 189-201.

[4] A.W. Bojanczyk, P. Van Dooren and G.H. Golub, *The periodic Schur decomposition. Algorithms and applications*, to appear in Proceedings SPIE, San Diego, July 1992.

[5] J. P. Charlier, M. Vanbegin and P. Van Dooren, *On efficient implementations of Kogbetliantz's algorithm for computing the singular value decomposition*, Numer. Math., 52 (1988), pp. 279–300.

[6] B. L. R. De Moor and G. H. Golub, *Generalized singular value decompositions: A proposal for a standardized nomenclature*, Manuscript NA-89-05, Numerical Analysis Project, Stanford University, Stanford, Calif., 1989.

[7] K. V. Fernando and S. J. Hammarling, *A product induced singular value decomposition for two matrices and balanced realisation*, in Linear Algebra in Signals, Systems and Control, B. N. Datta et al., Eds., SIAM, Philadelphia, Penn., 1988, pp. 128–140.

[8] M. T. Heath, A. J. Laub, C. C. Paige, and R. C. Ward, *Computing the SVD of a product of two matrices*, SIAM J. Sci. Statist. Comput., 7 (1986), pp. 1147–1159.

[9] C. C. Paige, *Computing the generalized singular value decomposition*, SIAM J. Sci. Statist. Comput., 7 (1986), pp. 1126–1146.

[10] H. Zha, *A numerical algorithm for computing the restricted SVD of matrix triplets*, to appear in Linear Algebra and Its Applications.

A parallel divide and conquer algorithm for the generalized real symmetric definite tridiagonal eigenproblem

Carlos F. Borges[*†] *and William B. Gragg*[‡§]

Abstract. We develop a parallel divide and conquer algorithm, by *extension*, for the generalized real symmetric definite tridiagonal eigenproblem. The algorithm employs techniques first proposed by Gu and Eisenstat to prevent loss of orthogonality in the computed eigenvectors for the *modification* algorithm. We examine numerical stability and adapt the insightful error analysis of Gu and Eisenstat to the arrow case. The algorithm incorporates an elegant zero finder with global monotone cubic convergence that has performed well in numerical experiments. A complete set of tested matlab routines implementing the algorithm is available on request from the authors.

Key words. generalized symmetric eigenvalue problem, divide and conquer algorithm.

AMS(MOS) subject classification. 65F15.

1 Introduction

We consider the problem of finding a matrix $U \in \Re^{n\times n}$ such that

$$U^T(T - S\lambda)U \equiv \Lambda - I\lambda,$$

is diagonal, or equivalently

$$U^T SU = I \quad \text{and} \quad U^T TU = \Lambda, \tag{1}$$

where

[*]Code Ma/Bc, Naval Postgraduate School, Monterey, CA 93943. Email: borges@waylon.math.nps.navy.mil.

[†]Research supported in part by the Naval Postgraduate School and the Interdisciplinary Project Center for Supercomputing at the ETH, Zürich.

[‡]Code Ma/Gr, Naval Postgraduate School, Monterey, CA 93943. Email: gragg@guinness.math.nps.navy.mil.

[§]Research supported in part by the Naval Postgraduate School.

$$T = \begin{bmatrix} \alpha_1 & \beta_1 & & & \\ \beta_1 & \alpha_2 & \beta_2 & & \\ & \beta_2 & & \ddots & \\ & & \ddots & & \beta_{n-1} \\ & & & \beta_{n-1} & \alpha_n \end{bmatrix}, \qquad S = \begin{bmatrix} \delta_1 & \gamma_1 & & & \\ \gamma_1 & \delta_2 & \gamma_2 & & \\ & \gamma_2 & & \ddots & \\ & & \ddots & & \gamma_{n-1} \\ & & & \gamma_{n-1} & \delta_n \end{bmatrix},$$

and S is assumed to be positive definite. This generalized eigenvalue problem has two special cases that are of interest in themselves. They are:

1. $S = I$, the ordinary tridiagonal eigenproblem.

2. $S = I$ and $\alpha_k \equiv 0$, the bidiagonal singular value problem (BSVP), by *perfect shuffle* of the Jordan matrix

$$\begin{bmatrix} 0 & B^T \\ B & 0 \end{bmatrix}$$

with B upper bidiagonal [16].

There are two phases to the divide and conquer algorithm, the divide (or split) phase, and the conquer (or consolidate) phase. We shall address these in order.

2 The algorithm

2.1 The divide phase

Denote by $\mathbf{e}_i$ the ith axis vector where the dimension will be clear from the context. Let $s, 1 \leq s \leq n$, be an integer, the *split index*, and consider the following block forms:

$$\begin{aligned} T &= \begin{bmatrix} T_1 & \mathbf{e}_{s-1}\beta_{s-1} & \\ \beta_{s-1}\mathbf{e}_{s-1}^T & \alpha_s & \beta_s\mathbf{e}_1^T \\ & \mathbf{e}_1\beta_s & T_2 \end{bmatrix}, \\ S &= \begin{bmatrix} S_1 & \mathbf{e}_{s-1}\gamma_{s-1} & \\ \gamma_{s-1}\mathbf{e}_{s-1}^T & \delta_s & \gamma_s\mathbf{e}_1^T \\ & \mathbf{e}_1\gamma_s & S_2 \end{bmatrix}. \end{aligned}$$

Note that $s = n$ is possible; then T_2, S_2, and $\mathbf{e}_1$ are empty [9, 10]. Suppose we solve the subproblems

$$U_k^T (T_k - S_k\lambda) U_k \equiv \Lambda_k - I\lambda \qquad (k = 1, 2). \tag{2}$$

The form of the subproblems is preserved. In particular, the matrices S_k are positive definite and, if T has a zero diagonal, so do the matrices T_k. Let

$$\hat{U} = \begin{bmatrix} U_1 & & \\ & 1 & \\ & & U_2 \end{bmatrix}.$$

Then

$$\hat{U}^T (T - S\lambda) \hat{U} = \begin{bmatrix} U_1^T(T_1 - S_1\lambda)U_1 & U_1^T\mathbf{e}_{s-1}(\beta_{s-1} - \gamma_{s-1}\lambda) & \\ (\beta_{s-1} - \gamma_{s-1}\lambda)\mathbf{e}_{s-1}^T U_1 & \alpha_s - \delta_s\lambda & (\beta_s - \gamma_s\lambda)\mathbf{e}_1^T U_2 \\ & U_2^T\mathbf{e}_1(\beta_s - \gamma_s\lambda) & U_2^T(T_2 - S_2\lambda)U_2 \end{bmatrix}.$$

2.2 The conquer phase

The conquer phase consists of solving the subproblems (2) from the divide phase, consolidating the solutions, and finally, solving the consolidated problem. Let

$$\mathbf{u}_1 = U_1^T\mathbf{e}_{s-1}, \qquad \mathbf{u}_2 = U_2^T\mathbf{e}_1,$$

where the U_k are solutions to (2). Then

$$\hat{U}^T (T - S\lambda) \hat{U} = \begin{bmatrix} \Lambda_1 - I\lambda & \mathbf{u}_1(\beta_{s-1} - \gamma_{s-1}\lambda) & \\ (\beta_{s-1} - \gamma_{s-1}\lambda)\mathbf{u}_1^T & \alpha_s - \delta_s\lambda & (\beta_s - \gamma_s\lambda)\mathbf{u}_2^T \\ & \mathbf{u}_2(\beta_s - \gamma_s\lambda) & \Lambda_2 - I\lambda \end{bmatrix}.$$

The right side is the sum of a diagonal and a *Swiss cross*:

$$\hat{U}^T (T - S\lambda) \hat{U} = \begin{bmatrix} x & & & & \\ & x & & & \\ & & x & & \\ & & & x & \\ & & & & x \end{bmatrix} + \begin{bmatrix} & & x & & \\ & & x & & \\ x & x & x & x & x \\ & & x & & \\ & & x & & \end{bmatrix}.$$

This can be permuted to an arrow matrix by a permutation similarity transformation with $P_s = [\mathbf{e}_1, \mathbf{e}_2, ..., \mathbf{e}_{s-1}, \mathbf{e}_{s+1}, ..., \mathbf{e}_n, \mathbf{e}_s]$. Thus

$$\begin{aligned} \hat{A}(\lambda) &:= P_s^T \hat{U}^T (T - S\lambda) \hat{U} P_s \\ &= \begin{bmatrix} \Lambda_1 & & \mathbf{u}_1\beta_{s-1} \\ & \Lambda_2 & \mathbf{u}_2\beta_s \\ \beta_{s-1}\mathbf{u}_1^T & \beta_s\mathbf{u}_2^T & \alpha_s \end{bmatrix} - \begin{bmatrix} I & & \mathbf{u}_1\gamma_{s-1} \\ & I & \mathbf{u}_2\gamma_s \\ \gamma_{s-1}\mathbf{u}_1^T & \gamma_s\mathbf{u}_2^T & \delta_s \end{bmatrix} \lambda \\ &= \begin{bmatrix} D & B\mathbf{u} \\ \mathbf{u}^T B & \alpha \end{bmatrix} - \begin{bmatrix} I & C\mathbf{u} \\ \mathbf{u}^T C & \gamma \end{bmatrix} \lambda, \end{aligned}$$

with

$$\mathbf{u} = \begin{bmatrix} \mathbf{u}_1 \\ \mathbf{u}_2 \end{bmatrix},$$

$$B = \begin{bmatrix} \beta_{s-1} I & \\ & \beta_s I \end{bmatrix}, \quad C = \begin{bmatrix} \gamma_{s-1} I & \\ & \gamma_s I \end{bmatrix}.$$

Since S and $\begin{bmatrix} I & C\mathbf{u} \\ \mathbf{u}^T C & \gamma \end{bmatrix}$ are congruent the latter inherits positive definiteness from the former. Its Cholesky decomposition is

$$\begin{bmatrix} I & C\mathbf{u} \\ \mathbf{u}^T C & \gamma \end{bmatrix} = \begin{bmatrix} I & \\ \mathbf{u}^T C & \rho \end{bmatrix} \begin{bmatrix} I & C\mathbf{u} \\ & \rho \end{bmatrix} = R^T R,$$

with $\rho^2 = \gamma - \mathbf{u}^T C^2 \mathbf{u} > 0$ the Schur complement in S of

$$\begin{bmatrix} S_1 & \\ & S_2 \end{bmatrix}.$$

Now

$$R^{-1} = \begin{bmatrix} I & -C\mathbf{u}/\rho \\ & 1/\rho \end{bmatrix}$$

and a second congruence transformation with R^{-1} give

$$\begin{aligned} A(\lambda) &:= R^{-T} \hat{A}(\lambda) R^{-1} \\ &= R^{-T} \begin{bmatrix} D & B\mathbf{u} \\ \mathbf{u}^T B & \alpha \end{bmatrix} R^{-1} - I\lambda \\ &= \begin{bmatrix} D & \mathbf{w} \\ \mathbf{w}^T & \omega \end{bmatrix} - I\lambda \quad =: \quad A - \lambda I \end{aligned}$$

with

$$\begin{aligned} \mathbf{w} &= \frac{(B - DC)\,\mathbf{u}}{\rho}, \\ \omega &= \frac{\alpha_s - \mathbf{u}^T\,(2B - DC)\,C\mathbf{u}}{\rho^2}. \end{aligned}$$

We have reduced the conquer step to the problem of solving an ordinary eigenproblem for a symmetric arrow matrix. If V is an orthogonal matrix with

$$AV = V\Lambda$$

and Λ diagonal, then (1) holds with

$$\begin{aligned} U &= \hat{U} P_s R^{-1} V \\ &= \begin{bmatrix} U_1 & & -U_1 \mathbf{u}_1 \gamma_{s-1}/\rho \\ & & 1/\rho \\ & U_2 & -U_2 \mathbf{u}_2 \gamma_s/\rho \end{bmatrix} V. \end{aligned}$$

It is useful that $\mathbf{v}_k = U_k\mathbf{u}_k$ can be computed in $O(n)$ time by solving $S_1\mathbf{v}_1 = \mathbf{e}_{s-1}$ and $S_2\mathbf{v}_2 = \mathbf{e}_1$ using the Cholesky factorization $S_1 = L_1L_1^T$ and the reverse Cholesky factorization $S_2 = L_2^TL_2$. In the case that only the eigenvalues are wanted it is only necessary to compute the first and last rows of the U-matrices which constitutes a further savings.

In summary, the conquer phase proceeds by consolidating the subproblems and building a full eigenproblem for an arrow matrix.

3 Solving the eigenproblem for the arrow

In this section we consider the solution of the eigenproblem for a real symmetric arrow matrix

$$A = \begin{bmatrix} D & \mathbf{b} \\ \mathbf{b}^T & \gamma \end{bmatrix}$$

where $A \in \Re^{n\times n}$ is symmetric, $D = \text{diag}(\mathbf{a})$, $\mathbf{a} = [\alpha_1, ..., \alpha_{n-1}]^T$, $\alpha_1 \geq \alpha_2 \geq ... \geq \alpha_{n-1}$, and $\mathbf{b} = [\beta_1, ..., \beta_{n-1}]^T \geq 0$. When A arises from the BSVD then $\mathbf{a}$ is odd and $\mathbf{b}$ is even, that is $\mathbf{a} + J\mathbf{a} = \mathbf{0}$ and $\mathbf{b} = J\mathbf{b}$, with J the *counter-identity*, the identity matrix with its columns reversed, and $\gamma = 0$.

If any $\beta_j = 0$ then it is possible to set $\lambda_j = \alpha_j$ and deflate the matrix since $\mathbf{e}_j$ is clearly an eigenvector [28]. We shall call this β-deflation and note that if $\beta_j < tol_\beta\|\mathbf{b}\|$ where tol_β is a small tolerance then a *numerical* deflation occurs. We derive a precise value for tol_β in section 4.4.

A second type of deflation occurs if applying a 2×2 rotation similarity transformation in the $(j, j+1)$-plane that takes β_j to zero introduces a sufficiently small element in the $(j, j+1)$ position of the matrix. This will be called a *combo*-deflation (see [15]). At each consolidation step we perform a sweep to check for β-deflations followed by a sweep to check for *combo*-deflations. The *combo*-deflation can be arranged so that the ordering of the α_j is preserved whenever one occurs. After deflation the new $\beta_{j+1}^* := \sqrt{\beta_j^2 + \beta_{j+1}^2} \geq \beta_{j+1}$ and hence no further β-deflation can occur. The *combo*-deflations can be disposed of with a single pass by backing up a single element whenever one occurs. Note that deflation is backward stable since it results in *small* backward errors in A. Deflation for the BSVD is more delicate involving a simultaneous sweep from both ends of the matrix. Care must be exercised at the center of the matrix.

After deflation the resulting matrix can be taken to have all $\beta_j > 0$ and the elements of the arrow *shaft* distinct and ordered, that is $\alpha_1 > \alpha_2 > ... > \alpha_{n-1}$. An arrow matrix of this form will be called *ordered* and *reduced*. Henceforth, we shall assume A is of this form.

The block Gauss factorization of $A - \lambda I$ is

$$A - \lambda I = \begin{bmatrix} I & 0 \\ \mathbf{b}^T(D-\lambda I)^{-1} & 1 \end{bmatrix} \begin{bmatrix} D-\lambda I & \mathbf{b} \\ 0^T & -f(\lambda) \end{bmatrix}$$

where f, the *spectral function* of A, is given by

$$f(\lambda) = \lambda - \gamma + \sum_{j=1}^{n-1} \frac{\beta_j^2}{\alpha_j - \lambda}.$$

This is a rational Pick function with a pole at infinity [1]. The *most general* form of a rational Pick function is

$$f(\lambda) = \delta\lambda - \gamma + \sum_{j=1}^{n-1} \frac{\beta_j^2}{\alpha_j - \lambda}, \qquad \delta \geq 0. \tag{3}$$

In relation to the various divide and conquer schemes, the case $\delta > 0$ corresponds with *extension*, $\delta = 0$ with *modification*, and $\delta = \gamma = 0$ with *restriction* [7].

Inspection of the graph of the spectral function reveals that the elements of the shaft interlace the eigenvalues

$$\lambda_1 > \alpha_1 > \lambda_2 > \ldots > \alpha_{n-1} > \lambda_n. \tag{4}$$

Moreover, in the present case, the derivative of the spectral function is bounded below by one so that its zeros are, in a certain sense, well determined.

3.1 The zero finder

The fundamental problem in finding the eigenvalues of an arrow is that of providing a stable and efficient method for finding the zeros of the spectral function. We now examine this problem in some detail.

The zero finding algorithm we present is globally convergent in the sense that the iteration will converge to the unique zero of f in (α_k, α_{k-1}) from any starting value in the closed interval $[\alpha_k, \alpha_{k-1}]$, where we put $\alpha_0 = +\infty$ and $\alpha_n = -\infty$. The zero finder converges monotonically at a *cubic* rate and applies to a general Pick function as given in formula (3).

3.2 Interior intervals

The iterative procedure for finding the unique zero of f in one of the interior intervals (α_k, α_{k-1}) proceeds as follows. Let x_0, $\alpha_k < x_0 < \alpha_{k-1}$, be an initial approximation to λ_k. If x_j is known choose

$$\phi_j(x) = \sigma + \frac{\omega_0}{\alpha_{k-1} - x} + \frac{\omega_1}{\alpha_k - x}$$

so that

$$\phi_j^{(i)}(x_j) = f^{(i)}(x_j), \quad i = 0, 1, 2. \tag{5}$$

Thus σ, ω_0, and ω_1 must satisfy

$$\begin{bmatrix} 1 & (\alpha_{k-1} - x_j)^{-1} & (\alpha_k - x_j)^{-1} \\ 0 & (\alpha_{k-1} - x_j)^{-2} & (\alpha_k - x_j)^{-2} \\ 0 & (\alpha_{k-1} - x_j)^{-3} & (\alpha_k - x_j)^{-3} \end{bmatrix} \begin{bmatrix} \sigma \\ \omega_0 \\ \omega_1 \end{bmatrix} = \begin{bmatrix} f(x_j) \\ f'(x_j) \\ f''(x_j) \end{bmatrix}$$

and we find

$$\begin{aligned} \sigma &= 3x_j - (\gamma + \alpha_{k-1} + \alpha_k) + \sum_{i \neq k-1,k} \frac{\beta_i^2}{\alpha_i - x_j} \frac{\alpha_i - \alpha_{k-1}}{\alpha_i - x_j} \frac{\alpha_i - \alpha_k}{\alpha_i - x_j}, \\ \omega_0 &= \beta_{k-1}^2 + \frac{(\alpha_{k-1} - x_j)^3}{\alpha_{k-1} - \alpha_k} \left(1 + \sum_{i \neq k-1,k} \frac{\beta_i^2}{(\alpha_i - x_j)^2} \frac{\alpha_i - \alpha_k}{\alpha_i - x_j} \right), \\ \omega_1 &= \beta_k^2 + \frac{(x_j - \alpha_k)^3}{\alpha_{k-1} - \alpha_k} \left(1 + \sum_{i \neq k-1,k} \frac{\beta_i^2}{(\alpha_i - x_j)^2} \frac{\alpha_i - \alpha_{k-1}}{\alpha_i - x_j} \right). \end{aligned}$$

Since $\omega_0 > 0$ and $\omega_1 > 0$ it follows that ϕ_j is a Pick function. Thus ϕ_j has a unique zero $x_{j+1} \in (\alpha_k, \alpha_{k-1})$. Also

$$\omega_0 > \beta_{k-1}^2 > 0 \quad , \qquad \omega_1 > \beta_k^2 > 0.$$

The error function

$$f(x) - \phi(x) = x - (\gamma + \sigma) + \sum_{i \neq k-1,k} \frac{\beta_i^2}{\alpha_i - x} + \frac{\beta_{k-1}^2 - \omega_0}{\alpha_{k-1} - x} + \frac{\beta_k^2 - \omega_1}{\alpha_k - x}$$

has n zeros, counting multiplicities. There are $n-3$ zeros exterior to (α_k, α_{k-1}) and three more at x_j. Thus the error function crosses zero exactly once in the interval (α_k, α_{k-1}). Hence x_{j+1} lies between x_j and λ_k, and the iteration is monotonically convergent from any starting guess $x_0 \in [\alpha_k, \alpha_{k-1}]$ as claimed. The cubic rate of convergence follows from (5).

Successive iterates can be found by solving quadratic equations. Rather than solve $\phi_j(x) = 0$ for x_{j+1} it is better to solve

$$\phi_j(x_j - \Delta) = 0$$

for the *increment* $\Delta = x_j - x_{j+1}$. Some rearrangement using (5) reduces this to

$$\alpha\Delta^2 + \beta\Delta - f = 0, \tag{6}$$

with

$$\alpha = \frac{\sigma}{(\alpha_{k-1} - x_j)(x_j - \alpha_k)}, \tag{7}$$

$$\beta = f'(x_j) - \left(\frac{1}{\alpha_{k-1} - x_j} + \frac{1}{\alpha_k - x_j} \right) f(x_j). \tag{8}$$

When shifts of the origin to the nearest pole [15] are used then one of α_{k-1} or α_k is zero. The computation of $\beta = \beta(x_j)$ should account for the fact that it has only *simple* poles at α_{k-1} and α_k.

If we start at the midpoint of the interval, $x_0 = (\alpha_{k-1} + \alpha_k)/2$, then we always have $\beta = \beta(x_j) \geq f'(x_j) \geq 1$. This can be seen by noting that $\beta(x_0) = f'(x_0)$ and that when $x_0 > \lambda_k$ then for all of the succeeding iterates $f(x_j) > 0$, by monotonicity, and $\frac{1}{\alpha_{k-1} - x_j} + \frac{1}{\alpha_k - x_j}$ is negative. If $x_0 < \lambda_k$ a similar argument applies. It follows that the increment can *always* be computed stably as

$$\Delta = \frac{2f/\beta}{1 + \sqrt{1 + \frac{2\alpha}{\beta}\frac{2f}{\beta}}}. \tag{9}$$

3.3 Exterior intervals

The treatment of the two exterior intervals is *geometrically* the same as above. Again, the approximating function has poles at the endpoints and the residues at these poles, and the constant term, are chosen to satisfy (5). We present the case for the interval (α_1, ∞), the case for the other exterior interval being similar. Now

$$\phi_j(x) = \omega_0 x - \sigma + \frac{\omega_1}{\alpha_1 - x}$$

with

$$\omega_0 = 1 + \sum_{i=2}^{n-1} \frac{\beta_i^2}{(x_j - \alpha_i)^2} \frac{\alpha_1 - \alpha_i}{x_j - \alpha_i} \geq 1,$$

$$\omega_1 = \beta_1^2 + \sum_{i=2}^{n-1} \beta_i^2 \left(\frac{x_j - \alpha_1}{x - \alpha_i} \right)^3 \geq \beta_1^2.$$

The inequalities are strict unless $n = 2$. Again we find (6) where now

$$\alpha = -\frac{1 + \sum_{i=2}^{n-1} \frac{\beta_i^2}{(x_j - \alpha_i)^2} \frac{\alpha_1 - \alpha_i}{x_j - \alpha_i}}{x_j - \alpha_1},$$

$$\beta = f'(x_j) + \frac{f(x_j)}{x_j - \alpha_1}.$$

These are limiting cases of (7) and (8) (introduce another pole $\alpha_0 > \alpha_1$ and let $\alpha_0 \to +\infty$). If $x_0 > \lambda_1$ then $f(\lambda) > 0$ so $\beta > f' > 1$ and Δ is again computed stably using (9). We obtain global monotone cubic convergence as before.

Contrary to the algorithms of [11, 12, 15] our algorithm is well-defined when starting at the endpoints of the intervals. The algorithm of [23] can start at the endpoints but has only quadratic convergence.

To guarantee that $x_0 \geq \lambda_1$ we take x_0 to be the iterate in $(\alpha_1, +\infty)$ from $+\infty$. As $x_0 \to +\infty$ the approximate Pick function tends to

$$\phi(x) = x - \gamma + \frac{\|\mathbf{b}\|^2}{\alpha_1 - x}. \tag{10}$$

Our *actual* starting guess is the zero of (10) in $(\alpha_1, +\infty)$:

$$x_0 = \begin{cases} \alpha_1 + \frac{\gamma-\alpha_1}{2} + \sqrt{\left(\frac{\gamma-\alpha_1}{2}\right)^2 + \|\mathbf{b}\|^2} & , \quad \gamma > \alpha_1, \\ \alpha_1 + \dfrac{\|\mathbf{b}\|^2}{\frac{\alpha_1-\gamma}{2} + \sqrt{\left(\frac{\gamma-\alpha_1}{2}\right)^2 + \|\mathbf{b}\|^2}} & , \quad \gamma \leq \alpha_1. \end{cases}$$

When shifts are used we have $\alpha_1 = 0$.

3.4 Orthogonality of the eigenvectors

It is essential that the computed eigenvectors of the arrow matrix be numerically orthogonal. As a point of entry into the further analysis of the algorithm we now examine the orthogonality of the eigenvectors following [15].

Consider the divided difference

$$\begin{aligned} f(\lambda, \mu) &= \frac{f(\lambda) - f(\mu)}{\lambda - \mu} = 1 + \sum_{j=1}^{n-1} \frac{\beta_j^2}{(\alpha_j - \lambda)(\alpha_j - \mu)} \\ &= 1 + \mathbf{b}^T (D - \lambda I)^{-1} (D - \mu I)^{-1} \mathbf{b}. \end{aligned} \tag{11}$$

Note that $\mu = \lambda$ gives $f'(\lambda) = 1 + \|(D - \lambda I)^{-1}\mathbf{b}\|_2^2$. If $f(\lambda) = 0$ then

$$v(\lambda) = \begin{bmatrix} \frac{\beta_j}{\lambda - \alpha_j} \\ 1 \end{bmatrix} = \begin{bmatrix} (\lambda I - D)^{-1}\mathbf{b} \\ 1 \end{bmatrix}$$

is an eigenvector of the arrow matrix $A = \begin{bmatrix} D & \mathbf{b} \\ \mathbf{b}^T & \gamma \end{bmatrix}$ associated with the eigenvalue λ, and

$$u(\lambda) = \frac{v(\lambda)}{\sqrt{f'(\lambda)}}$$

is the normalized eigenvector whose last element is positive. The ordering of A implies that its matrix of eigenvectors can be taken positive below and on the diagonal, and negative above.

Let $f(\lambda_0) = f(\mu_0) = 0$ with $\lambda_0 \neq \mu_0$. Thus λ_0 and μ_0 are distinct eigenvalues of A. The eigenvectors $u(\lambda_0)$ and $u(\mu_0)$ are orthonormal:

$$u(\lambda_0)^T u(\mu_0) = f(\lambda_0, \mu_0) = 0.$$

Let λ and μ be approximate eigenvalues in the sense that

$$\begin{aligned} -\delta_j &= \frac{\lambda - \lambda_0}{\alpha_j - \lambda_0}, \qquad |\delta_j| \le \frac{\delta}{1+\delta}, \\ -\delta_j' &= \frac{\mu - \mu_0}{\alpha_j - \mu_0}, \qquad |\delta_j'| \le \frac{\delta}{1+\delta}. \end{aligned} \tag{12}$$

Here $\delta > 0$ is hopefully, but not necessarily, close to the machine unit ϵ. Note that (12) is equivalent with

$$\frac{\alpha_j - \lambda}{\alpha_j - \lambda_0} = 1 + \delta_j, \quad \frac{\alpha_j - \mu}{\alpha_j - \mu_0} = 1 + \delta_j'.$$

These conditions imply that the approximate eigenvectors $u(\lambda)$ and $u(\mu)$ are nearly orthogonal. For we have

$$\begin{aligned} \sqrt{f'(\lambda) f'(\mu)}\, u(\lambda)^T u(\mu) &= f(\lambda, \mu) - f(\lambda_0, \mu_0) \\ &= \sum_{j=1}^{n-1} \frac{\beta_j^2}{(\alpha_j - \lambda)(\alpha_j - \mu)} \left(1 - \frac{(\alpha_j - \lambda)(\alpha_j - \mu)}{(\alpha_j - \lambda_0)(\alpha_j - \mu_0)}\right) \\ &= \sum_{j=1}^{n-1} \frac{\beta_j^2}{(\alpha_j - \lambda)(\alpha_j - \mu)} \left(\delta_j + \delta_j' + \delta_j \delta_j'\right). \end{aligned}$$

Since

$$|\delta_j + \delta_j' + \delta_j \delta_j'| \le \frac{2\delta}{1+\delta} + \frac{\delta^2}{(1+\delta)^2} \le 2\delta$$

then

$$\sqrt{f'(\lambda) f'(\mu)}\, u(\lambda)^T u(\mu) = 2\delta \mathbf{b}^T (D - \lambda I)^{-1} \Theta (D - \mu I)^{-1} \mathbf{b}$$

with $|\Theta| \le I$. Thus

$$\sqrt{f'(\lambda) f'(\mu)}\, |u(\lambda)^T u(\mu)| = 2\delta \|(D - \lambda I)^{-1} \mathbf{b}\|_2 \|(D - \mu I)^{-1} \mathbf{b}\|_2,$$

and so

$$|u(\lambda)^T u(\mu)| < 2\delta.$$

Condition (12) is stringent. If we let $\beta_k \to 0$ then it is easy to show that A can have an eigenvalue $\lambda_0 = \lambda_0(\beta_k) = \alpha_k + O(\beta_k^2)$; (12) then requires that the approximate eigenvalue λ satisfies a bound

$$|\lambda - \lambda_0| \leq O(\delta\beta_k^2),$$

which is difficult if $\beta_k/\|\mathbf{b}\|$ is only somewhat larger than machine precision, say $\epsilon^{3/4}$. Two techniques are used to attempt to satisfy (12) – shifts of the origin [15], and simulated extended precision (SEP) arithmetic [26, 14]. Condition (12) means that

$$|\lambda - \lambda_0| < \delta \min\{\lambda_0 - \alpha_k, \alpha_{k-1} - \lambda_0\}.$$

When shifts are used it means that λ is nearly $fl(\lambda_0)$.

4 Numerical stability of the algorithm

We now give a partial analysis of the stability of this approach to the eigenproblem for the symmetric arrow matrix. Observe that

$$f(\lambda) = \frac{p(\lambda)}{q(\lambda)} := \frac{\prod_{j=1}^{n}(\lambda - \lambda_j)}{\prod_{j=1}^{n-1}(\lambda - \alpha_j)}.$$

The following *inverse eigenvalue problem* [6] is important: given $\{\alpha_j\}$ and $\{\lambda_j\}$ satisfying (4), find $\{\beta_j\}$ and γ so that $\lambda(A) = \{\lambda_j\}$. This problem is simply solved by computing the residues of the partial fraction decomposition of f. In particular

$$\begin{aligned}
\beta_k^2 &= \lim_{\lambda \to \alpha_k} (\alpha_k - \lambda)\frac{p(\lambda)}{q(\lambda)} \\
&= -\frac{\prod_{j=1}^{n}(\alpha_k - \lambda_j)}{\prod_{j \neq k}(\alpha_k - \alpha_j)}, \\
\gamma &= \sum_{j=1}^{n} \lambda_j - \sum_{j=1}^{n-1} \alpha_j.
\end{aligned}$$

For fixed $\{\alpha_j\}$, the elements of the arrow *head*, $\{\beta_k\}$ and γ, are *explicitly known* functions of the eigenvalues.

Now let $\{\hat{\lambda}_j\}$ be a set of *approximate* eigenvalues of A satisfying (4). Then

$$\hat{\beta}_k^2 = -\frac{\prod_{j=1}^{n}(\alpha_k - \hat{\lambda}_j)}{\prod_{j \neq k}(\alpha_k - \alpha_j)} \qquad (\hat{\beta}_k > 0), \tag{13}$$

$$\hat{\gamma} = \sum_{j=1}^{n} \hat{\lambda}_j - \sum_{j=1}^{n-1} \alpha_j, \tag{14}$$

define a *modified* matrix $\hat{A}$ with $\lambda(\hat{A}) = \{\hat{\lambda}_j\}$. To obtain a backward error analysis for the complete eigen*value* problem we bound the differences $\hat{\beta}_k - \beta_k$ and $\hat{\gamma} - \gamma$.

4.1 Error analysis for the Dongarra-Sorensen condition

We give an error analysis using the Dongarra-Sorensen condition

$$-\frac{\hat{\lambda}_j - \lambda_j}{\alpha_k - \lambda_j} = \delta_{j,k}, \quad |\delta_{j,k}| \leq \delta, \tag{15}$$

where $\delta = O(\epsilon)$ is of the order of the machine unit, simplifying that in [6].

Rearrangement of (15) gives

$$\hat{\lambda}_j - \alpha_k = (\lambda_j - \alpha_k)(1 + \delta_{j,k}).$$

It follows that

$$\hat{\beta}_k^2 = \beta_k^2 \prod_{j=1}^{n} (1 + \delta_{j,k}) = \beta_k^2 \left(1 + \sum_{j=1}^{n} \delta'_{j,k}\right),$$

and

$$\hat{\beta}_k = \beta_k \left(1 + \frac{1}{2} \sum_{j=1}^{n} \delta''_{j,k}\right),$$

with the $\delta'_{j,k}$ and $\delta''_{j,k}$ at most only slightly larger than the $\delta_{j,k}$. Thus

$$\left| \frac{\hat{\beta}_k - \beta_k}{\beta_k} \right| \leq \frac{n}{2} \delta'',$$

where $\delta'' = O(\epsilon)$ is only slightly larger than δ.

Now (14) becomes

$$\hat{\gamma} = \gamma + \sum_{j=1}^{n} (\lambda_j - \alpha_{k(j)}) \delta_{j,k(j)}$$

with $\alpha_{k(j)}$ one of the poles of f. Thus

$$|\hat{\gamma} - \gamma| \leq \delta \sum_{j=1}^{n} |\lambda_j - \alpha_{k(j)}|.$$

To minimize this bound we choose $\alpha_{k(j)}$ to be a pole of f closest to λ_j. Clearly, $\alpha_{k(1)} = \alpha_1$ and $\alpha_{k(n)} = \alpha_{n-1}$, so

$$|\hat{\gamma} - \gamma| \leq \delta \left((\lambda_1 - \alpha_1) + \sum_{j=2}^{n-1} |\lambda_j - \alpha_{k(j)}| + (\alpha_{n-1} - \lambda_n) \right).$$

For $1 < j < n$ a closest pole to λ_j is either α_j or α_{j-1}. The distance

$$|\lambda_j - \alpha_{k(j)}| = \min \{\lambda_j - \alpha_j, \alpha_{j-1} - \lambda_j\}$$

is maximized when λ_j is the midpoint of the interval (α_j, α_{j-1}), and the value of the maximum is $(\alpha_j + \alpha_{j-1})/2$. Thus

$$\begin{aligned} |\hat{\gamma} - \gamma| &\leq \delta \left((\lambda_1 - \alpha_1) + \frac{1}{2} \sum_{j=2}^{n-1} (\alpha_{j-1} - \alpha_j) + (\alpha_{n-1} - \lambda_n) \right) \\ &= \delta \left((\lambda_1 - \lambda_n) - \frac{\alpha_1 - \alpha_{n-1}}{2} \right) \\ &\leq \delta(\lambda_1 - \lambda_n) \leq 2\delta \|A\|_2. \end{aligned}$$

In summary, the Dongarra-Sorensen condition implies small relative errors in each β_k and a small absolute error in γ. For the BSVD this implies small element-wise relative errors since the condition $\gamma = \hat{\gamma} = 0$ is enforced by $\hat{\lambda}_j + \hat{\lambda}_{n+1-j} = 0$ (only half of the eigenvalues are actually computed, the rest follow from this condition).

4.2 Rounding error analysis of the computation of $f(\lambda)$

The choice of a termination criterion depends on a careful rounding error analysis of the particular manner in which we compute $f(\lambda)$. Let $\{\alpha_i\}$, $\{\beta_i\}$, and γ be floating point numbers. We *represent* λ as the ordered pair of floating point numbers (σ, μ) where the *shift* σ is a pole closest to λ, and $\lambda := \sigma + \mu$. For the exterior intervals we have $\sigma = \alpha_1$ or $\sigma = \alpha_{n-1}$. For the interior intervals σ can be determined by evaluating f at the midpoint and checking the sign. We compute $f(\lambda)$ as

$$f_\sigma(\mu) = \sum_{j=1}^{n-1} \frac{\beta_j^2}{\alpha_j' - \mu} + (\mu - \gamma'),$$

with the standard operation precedence rules, where

$$\alpha_j' = \alpha_j - \sigma \quad \text{and} \quad \gamma' = \gamma - \sigma.$$

We use Wilkinson's notation: $fl(x * y) = (x * y)(1 + \delta)$ with $|\delta| \leq \epsilon/(1 + \epsilon)$ and $\epsilon = 2^{-t}$ the machine unit. More generally, ϵ denotes numbers not essentially larger than 2^{-t} [27] and the rounding errors δ satisfy $|\delta| < \epsilon$.

We define

$$fl(\alpha_j - \lambda) := fl(\alpha_j' - \mu) = fl((\alpha_j - \sigma) - \mu).$$

If $\sigma = \alpha_k$ then

$$fl(\alpha_k - \lambda) = -\mu = \alpha_k - \lambda,$$

with no rounding error. For $j \neq k$,

$$fl(\alpha_j - \lambda) = (\alpha_j - \lambda)\left(1 + \frac{\alpha_j - \alpha_k}{\alpha_j - \lambda}\delta + \delta'\right),$$

and since α_k is a pole *closest* to λ then $\left|\frac{\alpha_j - \alpha_k}{\alpha_j - \lambda}\right| \le 2$. Thus all terms $\alpha_j - \lambda$ are computed with small relative errors:

$$fl(\alpha_j - \lambda) = (\alpha_j - \lambda)(1 + 3\delta_j), \quad |\delta_j| < \epsilon. \tag{16}$$

When computing $f(\lambda) = f_\sigma(\mu)$ we add the term $\lambda - \gamma = (\lambda - \sigma) - (\gamma - \sigma)$ last. A routine error analysis using (16) and

$$|\lambda - \gamma| \le |f(\lambda)| + \sum_{j=1}^{n-1} \frac{\beta_j^2}{|\alpha_j - \lambda|}$$

to eliminate the term $|\lambda - \gamma|$ from the error bound gives

$$|fl(f(\lambda)) - f(\lambda)| \le \epsilon\left(3|f(\lambda)| + |\sigma - \lambda| + (n+5)\sum_{j=1}^{n-1} \frac{\beta_j^2}{|\alpha_j - \lambda|}\right)$$

which implies

$$|fl(f(\lambda))| \le (1 + 3\epsilon)|f(\lambda)| + \epsilon\left(|\sigma - \lambda| + (n+5)\sum_{j=1}^{n-1} \frac{\beta_j^2}{|\alpha_j - \lambda|}\right). \tag{17}$$

4.3 Termination

Our goal is to choose a termination criterion so that we stop when λ is as close to the true eigenvalue λ_k as possible. Let $\mu = \lambda_k$ in (11) with $f(\lambda_k) = 0$. Now $\alpha_k < \lambda_k < \alpha_{k-1}$. Also let $\alpha_k < \lambda < \alpha_{k-1}$. Then the terms $\alpha_j - \lambda$ and $\alpha_j - \lambda_k$ have the same sign and

$$|\lambda - \lambda_k| \le \frac{|f(\lambda)|}{1 + \sum_{j=1}^{n-1} \frac{\beta_j^2}{|\alpha_j - \lambda||\alpha_j - \lambda_k|}}. \tag{18}$$

To obtain an upper bound for $|\lambda - \lambda_k|$ we need an upper bound for $|f(\lambda)|$ and a lower bound for the denominator. For the latter we have

$$1 + \sum_{j=1}^{n-1} \frac{\beta_j^2}{|\alpha_j - \lambda||\alpha_j - \lambda_k|} \ge 1 + \frac{\sum_{j=1}^{n-1} \frac{\beta_j^2}{|\alpha_j - \lambda|}}{\max_j |\alpha_j - \lambda_k|}. \tag{19}$$

Let us determine how small $|f(\lambda)|$ is when λ is the *rounded representation* of λ_k. This is

$$\begin{aligned}\tilde{\lambda} &:= \sigma + fl(\mu_k) = \sigma + \mu_k(1+\delta) \\ &= \lambda_k + \mu_k\delta = \lambda_k + (\lambda_k - \sigma)\delta\end{aligned}$$

and we have

$$|\sigma - \lambda_k| = \min_j |\alpha_j - \lambda_k|.$$

Thus

$$\begin{aligned}|f(\tilde{\lambda})| &= |\tilde{\lambda} - \lambda_k| \left(1 + \sum_{j=1}^{n-1} \frac{\beta_j^2}{(\alpha_j - \tilde{\lambda})(\alpha_j - \lambda_k)}\right) \\ &= |(\sigma - \lambda_k)\delta| \left(1 + \sum_{j=1}^{n-1} \frac{\beta_j^2}{(\alpha_j - \tilde{\lambda})(\alpha_j - \lambda_k)}\right) \\ &\le \epsilon \left(|\sigma - \lambda_k| + \sum_{j=1}^{n-1} \frac{\beta_j^2}{|\alpha_j - \tilde{\lambda}|}\right).\end{aligned}$$

From (17),

$$|fl(f(\tilde{\lambda}))| \le \epsilon \left(|\sigma - \lambda_k| + |\tilde{\lambda} - \sigma| + (n+6) \sum_{j=1}^{n-1} \frac{\beta_j^2}{|\alpha_j - \tilde{\lambda}|}\right).$$

Since $\lambda_k - \sigma = (\tilde{\lambda} - \sigma)/(1+\delta)$ then

$$|fl(f(\tilde{\lambda}))| \le \epsilon \left(2|\tilde{\lambda} - \sigma| + (n+6) \sum_{j=1}^{n-1} \frac{\beta_j^2}{|\alpha_j - \tilde{\lambda}|}\right).$$

We terminate and set $\hat{\lambda}_k := \lambda$ when

$$|fl(f(\lambda))| \le 2\epsilon \left(2|\lambda - \sigma| + (n+6) \sum_{j=1}^{n-1} \frac{\beta_j^2}{|\alpha_j - \lambda|}\right).$$

Inequality (17) also holds if $f(\lambda)$ and $fl(f(\lambda))$ are interchanged. Thus

$$|f(\hat{\lambda}_k)| \le \epsilon \left(5|\hat{\lambda}_k - \sigma| + (3n+17) \sum_{j=1}^{n-1} \frac{\beta_j^2}{|\alpha_j - \hat{\lambda}_k|}\right). \tag{20}$$

From (18) and (19)

$$|\hat{\lambda}_k - \lambda_k| \le \epsilon \max_j |\lambda_k - \alpha_j| \frac{5|\sigma - \tilde{\lambda}_k| + (3n+17) \sum_{j=1}^{n-1} \frac{\beta_j^2}{|\alpha_j - \hat{\lambda}_k|}}{\max_j |\lambda_k - \alpha_j| + \sum_{j=1}^{n-1} \frac{\beta_j^2}{|\alpha_j - \hat{\lambda}_k|}}.$$

Since $|\sigma-\tilde{\lambda}_k| \leq |\sigma-\lambda_k|+|\tilde{\lambda}_k-\lambda_k|$ and $|\sigma-\lambda_k| \leq \max_j |\alpha_j-\lambda_k|$ the computed eigenvalues satisfy

$$\begin{aligned} |\hat{\lambda}_k-\lambda_k| &\leq (3n+17)\epsilon \max_j |\alpha_j-\lambda_k| \\ &\leq 6(n+6)\epsilon\|A\|_2. \end{aligned} \tag{21}$$

4.4 Error analysis for the Gu-Eisenstat condition

From $\hat{\gamma}-\gamma=\sum_{j=1}^{n}(\hat{\lambda}_j-\lambda_j)$ and (21) we find

$$|\hat{\gamma}-\gamma| \leq 6n(n+6)\epsilon\|A\|_2.$$

We have noted that the Dongarra-Sorensen condition (15) is stringent. It is natural to ask for small *absolute* errors in the β_k. If we replace $\delta_{j,k}$ by $\delta_{j,k}/\beta_k$ in the analysis in section 4.1 we find that

$$\hat{\beta}_k=\beta_k\left(1+\frac{1}{2}\sum_{j=1}^{n-1}\frac{\delta''_{j,k}}{\beta_k}\right)=\beta_k+\frac{1}{2}\sum_{j=1}^{n-1}\delta''_{j,k},$$

and

$$|\hat{\beta}_k-\beta_k| \leq \frac{n}{2}\delta'', \quad |\delta''| \leq \delta(1+O(\epsilon)),$$

are implied by the *Gu-Eisenstat condition*

$$-\beta_k\frac{\hat{\lambda}_j-\lambda_j}{\alpha_k-\lambda_j}=\delta_{j,k}, \quad |\delta_{j,k}| \leq \delta.$$

We must bound δ.

From (20)

$$|\hat{\lambda}_k-\lambda_k|\left(1+\sum_{j=1}^{n-1}\frac{\beta_j^2}{(\alpha_j-\hat{\lambda}_k)(\alpha_j-\lambda_k)}\right) \leq m\epsilon\left(|\hat{\lambda}_k-\sigma|+\sum_{j=1}^{n-1}\frac{\beta_j^2}{|\alpha_j-\hat{\lambda}_k|}\right)$$

with $m=3(n+6)$. Using

$$|\lambda-\sigma| \leq |\lambda-\lambda_k|+|\lambda_k-\sigma|$$

and the Gu-Eisenstat inequality,

$$\frac{1}{|\alpha_j-\lambda|} \leq \left|\frac{1}{(\alpha_j-\lambda)(\alpha_j-\lambda_k)}\right|^{1/2}+\frac{|\lambda_k-\lambda|}{(\alpha_j-\lambda)(\alpha_j-\lambda_k)},$$

we get

$$|\hat{\lambda}_k - \lambda_k| \le m\epsilon \left(|\lambda_k - \sigma| + \frac{\sum_{j=1}^{n-1} \frac{\beta_j^2}{[(\alpha_j - \hat{\lambda}_k)(\alpha_j - \lambda_k)]^{1/2}}}{\sum_{j=1}^{n-1} \frac{\beta_j^2}{(\alpha_j - \hat{\lambda}_k)(\alpha_j - \lambda_k)}} \right),$$

where ϵ has been increased to $\epsilon/(1 - m\epsilon)$.

By Cauchy's inequality,

$$\begin{aligned} |\hat{\lambda}_k - \lambda_k| &\le m\epsilon \left(|\lambda_k - \sigma| + \frac{\|\mathbf{b}\|}{\left(\sum_{j=1}^{n-1} \frac{\beta_j^2}{(\alpha_j - \hat{\lambda}_k)(\alpha_j - \lambda_k)}\right)^{1/2}} \right) \\ &\le m\epsilon \left(|\lambda_k - \sigma| + \frac{\|\mathbf{b}\|}{\beta_j} \left[(\alpha_j - \hat{\lambda}_k)(\alpha_j - \lambda_k) \right]^{1/2} \right), \end{aligned}$$

for every j. The arithmetic-geometric mean inequality and the triangle inequality yield

$$|\hat{\lambda}_k - \lambda_k| \le m\epsilon \left(|\lambda_k - \sigma| + \frac{\|\mathbf{b}\|}{\beta_j} \left(|\alpha_j - \lambda_k| + \frac{1}{2} |\lambda_k - \hat{\lambda}_k| \right) \right).$$

Thus

$$\begin{aligned} \left(1 - \frac{m\epsilon}{2} \frac{\|\mathbf{b}\|}{\beta_j}\right) |\hat{\lambda}_k - \lambda_k| &\le m\epsilon \left(|\lambda_k - \sigma| + |\alpha_j - \lambda_k| \frac{\|\mathbf{b}\|}{\beta_j} \right) \\ &= m\epsilon \|\mathbf{b}\| \frac{|\alpha_j - \lambda_k|}{\beta_j} \left(1 + \frac{|\lambda_k - \sigma|}{|\lambda_k - \alpha_j|} \frac{\beta_j}{\|\mathbf{b}\|} \right) \\ &\le 2m\epsilon \|\mathbf{b}\| \frac{|\alpha_j - \lambda_k|}{\beta_j}. \end{aligned}$$

If $m\epsilon\|\mathbf{b}\| \le \beta_j$ for all j, then

$$|\delta_{j,k}| \le 4m\epsilon\|\mathbf{b}\|.$$

and consequently

$$|\hat{\beta}_k - \beta_k| \le 6n(n+6)\epsilon\|\mathbf{b}\|.$$

Thus tol_β is $m\epsilon$. If $\beta_k < 3(n+6)\epsilon\|b\|$ we replace β_k by zero and accept α_k as an eigenvalue with normalized eigenvector $\mathbf{e}_k$.

The computed eigenvectors of A are taken to be those of the nearby matrix $\hat{A}$. Because of (13) and (16) they are computed to high relative precision elementwise and hence are numerically orthogonal [20].

Acknowledgements. We wish to thank Ming Gu and Stan Eisenstat for providing a preprint of their manuscript [20]. It is our understanding that they have independently extended their results to include the arrow case in [21].

References

[1] N. I. AKHIEZER, *The classical moment problem and some related questions in analysis*, Hafner Publishing Company, 1961.

[2] G. S. AMMAR, L. REICHEL, AND D. C. SORENSEN, *An implementation of a divide and conquer algorithm for the unitary eigenproblem*, ACM Trans. Math. Software, 18 (1992), pp. 292–307.

[3] P. ARBENZ, *Computing eigenvalues of banded symmetric Toeplitz matrices*, SIAM J. Sci. Statist. Comput., 12 (1991), pp. 743–754.

[4] ——, *Divide and conquer algorithms for the bandsymmetric eigenvalue problem*, in Parallel Computing '91, D. J. Evans, G. R. Joubert, and H. Liddell, eds., Elsevier Science Publishers B. V., Amsterdam, 1992, pp. 151–158.

[5] P. ARBENZ AND G. H. GOLUB, *On the spectral decomposition of hermitian matrices modified by low rank perturbations with applications*, SIAM J. Matrix Anal. Appl., 9 (1988), pp. 40–58.

[6] J. L. BARLOW, *Error analysis of update methods for the symmetric eigenvalue problem*, SIAM J. Matrix Anal. Appl., to appear.

[7] C. BEATTIE AND D. W. FOX, *Schur complements and the Weinstein-Aronszajn theory for modified matrix eigenvalue problems*, Linear Algebra Appl., 108 (1988), pp. 37–61.

[8] C. BEATTIE AND C. J. RIBBENS, *Parallel solution of a generalized symmetric matrix eigenvalue problem*, Working paper, Department of Mathematics, Virginia Polytechnic Institute and State University, Blacksburg, Virginia, 1992.

[9] A. A. BEEX AND M. P. FARGUES, *Highly parallel recursive/iterative Toeplitz eigenspace decomposition*, IEEE Trans. Acoustics, Speech, Signal Proc., 37 (1989), pp. 1765–1768.

[10] A. A. BEEX, D. M. WILKES, AND M. P. FARGUES, *The C-RISE algorithm and the generalized eigenvalue problem*, Signal Processing. Submitted.

[11] J. R. BUNCH AND C. NIELSEN, *Updating the singular value decomposition*, Numer. Math., 31 (1978), pp. 111–129.

[12] J. R. BUNCH, C. NIELSEN, AND D. C. SORENSEN, *Rank one modification of the symmetric eigenproblem*, Numer. Math., 31 (1978), pp. 31–48.

[13] J. J. M. CUPPEN, *A divide and conquer method for the symmetric tridiagonal eigenproblem*, Numer. Math., 36 (1981), pp. 177–195.

[14] T. J. DEKKER, *A floating-point technique for extending the available precision*, Numer. Math., 18 (1971), pp. 224–242.

[15] J. J. DONGARRA AND D. C. SORENSEN, *A fully parallel algorithm for the symmetric eigenvalue problem*, SIAM J. Sci. Statist. Comput., 8 (1987), pp. 139–154.

[16] G. Golub and W. Kahan, *Calculating the singular values and pseudo-inverse of a matrix*, J. Soc. Indust. Appl. Math. Ser. B Numer. Anal., 2 (1965).

[17] G. H. Golub, *Some modified matrix eigenvalue problems*, SIAM Rev., 15 (1973), pp. 318–334.

[18] W. B. Gragg and L. Reichel, *A divide and conquer method for unitary and orthogonal eigenproblems*, Numer. Math., 57 (1990), pp. 695–718.

[19] W. B. Gragg, J. R. Thornton, and D. D. Warner, *Parallel divide and conquer algorithms for the symmetric tridiagonal eigenproblem and bidiagonal singular value problem*, in Modeling and Simulation, vol. 23, part 1, W. G. Vogt and M. H. Mickle, eds., Univ. Pittsburgh School of Engineering, 1992, pp. 49–56.

[20] M. Gu and S. C. Eisenstat, *A stable and efficient algorithm for the rank-one modification of the symmetric eigenproblem*, SIAM J. Matrix Anal. Appl., to appear.

[21] M. Gu and S. C. Eisenstat, *A divide-and-conquer algorithm for the symmetric tridiagonal eigenproblem*, Working paper, Department of Computer Science, Yale University (1992).

[22] E. R. Jessup and D. C. Sorensen, *A parallel algorithm for computing the singular value decomposition of a matrix*, SIAM J. Matrix Anal. Appl., to appear.

[23] R.-C. Li, *Solving secular equations stably and efficiently*, Working paper, Department of Mathematics, University of California, Berkeley, (1992).

[24] K. Löwner, *Über monotone Matrixfunktionen*, Math Z., 38 (1934), pp. 177–216.

[25] D. P. O'Leary and G. W. Stewart, *Computing the eigenvalues and eigenvectors of arrowhead matrices*, J. Comp. Phys., 90 (1990), pp. 497–505.

[26] D. C. Sorensen and P. T. Tang, *On the orthogonality of eigenvectors computed by divide and conquer techniques*, SIAM J. Numer. Anal., 28 (1991), pp. 1752–1775.

[27] J. H. Wilkinson, *Rounding errors in algebraic processes*, Prentice-Hall, 1963.

[28] J. H. Wilkinson, *The algebraic eigenvalue problem*, Oxford University Press, 1965.

A parallel implementation of the GMRES method

Daniela Calvetti [*,†]
Johnny Petersen [‡]
Lothar Reichel [§,¶]

Abstract. The GMRES method by Saad and Schultz is one of the most popular iterative methods for solving large sparse nonsymmetric linear systems of equations. The method is usually implemented using the Arnoldi process, based on the modified Gram-Schmidt (MGS) method, to compute orthonormal bases of certain Krylov subspaces. The MGS method requires many vector-vector operations, which can be difficult to implement efficiently on vector and parallel computers due to the low granularity of these operations. In [3, 4] a new way to organize the computations in the GMRES method is presented, such that the vector-vector operations of the MGS method are replaced by the task of computing a QR factorization of a certain dense matrix. This paper presents timings for a parallel implementation on an IBM 3090-600S VF computer.

Key words. GMRES method, parallel implementation, Krylov space basis, Leja points.

AMS(MOS) subject classifications. 65F10, 65N22, 65Y05.

1 Introduction

The Generalized Minimal Residual (GMRES) method introduced by Saad and Schultz [16] is one of the most powerful iterative schemes for the numerical solution of large sparse nonsymmetric linear systems of equations

$$Ax = b, \tag{1.1}$$

*Department of Pure and Applied Mathematics,Stevens Institute of Technology, Hoboken, NJ 07030, USA.
†Research in part supported by IBM Bergen Scientific Centre.
‡IBM Bergen Scientific Centre, Thormøhlensgaten 55, N-5008 Bergen, Norway.
§Department of Mathematics and Computer Science, Kent State University, Kent, OH 44242, USA.
¶Research in part supported by IBM Bergen Scientific Centre and NSF grant DMS-9002884.

Numerical Linear Algebra

where $A \in \mathbb{R}^{N\times N}$ is nonsingular. In this method one chooses an initial approximate solution x_0, defines the residual vector $r_0 := b - Ax_0$, and then computes a better approximate solution $x_1 := x_0 + z_0 \in x_0 + \mathrm{K}_m(A, r_0)$, such that

$$\|b - Ax_1\| = \min_{z\in \mathrm{K}_m(A,r_0)} \|b - A(x_0 + z)\|, \tag{1.2}$$

where $\mathrm{K}_m(A, r_0) := \mathrm{span}\{r_0, Ar_0, \ldots, A^{m-1}r_0\}$ is a Krylov subspace and m is a given positive integer. Throughout this paper $\|\cdot\|$ denotes the Euclidean norm.

Saad and Schultz [16] solve the problem (1.2) by first computing an orthonormal basis $\{v_k\}_{k=1}^{m+1}$ of $\mathrm{K}_{m+1}(A, r_0)$, with $v_1 := r_0/\|r_0\|$, by the Arnoldi process, such that

$$AV_m = V_{m+1}H, \tag{1.3}$$

where $V_i := [v_1, v_2, \ldots, v_i]$, and $H = [\eta_{jk}] \in \mathbb{R}^{(m+1)\times m}$ is an essentially upper Hessenberg matrix. Substitution of (1.3) into (1.2) yields

$$\begin{aligned} \min_{z\in \mathrm{K}_m(A,r_0)} \|r_0 - Az\| &= \min_{y\in \mathrm{R}^m} \|r_0 - AV_m y\| = \min_{y\in \mathrm{R}^m} \|r_0 - V_{m+1}Hy\| \\ &= \min_{y\in \mathrm{R}^m} \| \, \|r_0\|e_1 - Hy\| , \end{aligned} \tag{1.4}$$

where the last equality follows from $V_{m+1}^T r_0 = \|r_0\|e_1$. Here and below e_j denotes the jth axis vector in $\mathbb{R}^{m+1}$. The solution y_0 of the least squares problem (1.4) is determined by the QR factorization of the matrix H, and we obtain a new approximate solution $x_1 := x_0 + V_m y_0$ of (1.1) and the corresponding residual vector $r_1 := b - Ax_1$. We remark that r_1 can also be evaluated according to $r_1 := r_0 - V_{m+1}Hy_0$, but numerical experiments by Karlsson [11] show that this formula can yield lower accuracy. It is therefore not used in the present paper. For numerical stability, the Arnoldi process is implemented using the modified Gram-Schmidt (MGS) method. This leads to the following algorithm.

Algorithm 1.1 (GMRES implementation by Saad and Schultz [16])
Input: m, x_0, $r_0 := b - Ax_0$;
Output: x_1, H; *(The computed* η_{jk} *are the nonvanishing entries of* H *in (1.3).)*
$v_1 := r_0/\|r_0\|$;
for $k := 1, 2, \ldots, m$ *do*
  $w := Av_k$;
  for $j := 1, 2, \ldots, k$ *do*
    $\eta_{jk} := w^T v_j$; $w := w - \eta_{jk}v_j$;
  end j;
  $\eta_{k+1,k} := \|w\|$; $v_{k+1} := w/\eta_{k+1,k}$;
end k;
Solve (1.4) for y_0 *by computing the QR factorization of* H;
$x_1 := x_0 + V_m y_0$; $r_1 := b - Ax_1$;

The storage requirement of Algorithm 1.1 grows linearly with m, and the number of arithmetic floating point operations (flops) required grows quadratically with m. Therefore, one generally chooses a fixed value of m, say $10 \le m \le 50$, and computes an approximate solution x_1 by Algorithm 1.1. If the norm of the residual error $r_1 := b - Ax_1$ is not sufficiently small, then one seeks to improve x_1 by solving a minimization problem analogous with (1.2). This gives rise to the cyclic GMRES algorithm, also introduced by Saad and Schultz [16].

Algorithm 1.2 (Cyclic GMRES(m) algorithm [16])
Input: m, x_0, $r_0 := b - Ax_0$, $\epsilon > 0$;
Output: *approximate solution* x_l *such that* $\|b - Ax_l\| \leq \epsilon$;
for $j := 0, 1, 2, \ldots$ *until* $\|r_j\| \leq \epsilon$ *do*

$$\textit{Solve} \quad \min_{z \in \mathrm{K}_m(A, r_j)} \|r_j - Az\| \quad \textit{for} \quad z_j \in \mathrm{K}_m(A, r_j); \tag{1.5}$$

$$\begin{aligned} x_{j+1} &:= x_j + z_j; \\ r_{j+1} &:= r_j - Az_j; \end{aligned} \tag{1.6}$$

end j;

The solution of (1.2) or (1.5) by the Arnoldi process gives rise to many vector-vector operations. These operations can be difficult to implement efficiently on vector and parallel computers due to their low granularity.

Bai et al. [3, 4] describe a scheme for the solution of (1.5), in which the vector-vector operations in Algorithm 1.1 are replaced by the task of computing the QR factorization of a dense $N \times (m+1)$ matrix. The algorithm for the computation of the QR factorization can be selected so as to be suitable for the computer at hand. In particular, the scheme in [3, 4] for the solution of (1.5) allows an implementation of Algorithm 1.2, with higher granularity and roughly the same number of flops, than if (1.5) were solved using Algorithm 1.1. Timings for a sequential computer (an IBM RISC System/6000 workstation) in [3, 4] confirm that an implementation of Algorithm 1.2 based on the scheme for the solution of (1.5) presented in [3, 4] is competitive; see also Section 3 for a discussion.

The purpose of the this paper is to present a parallel implementation of the scheme described in [3, 4], and to show timings for an implementation on an IBM 3090-600S VF computer.

The scheme in [3, 4] and in the present paper generalizes an algorithm by Hindmarsh and Walker [9], which represents the Krylov subspaces by using scaled monomial bases, i.e., $\mathrm{K}_{m+1}(A, r_j) = \mathrm{span}\{\sigma_0 r_j, \sigma_1 A r_j, \ldots, \sigma_m A^m r_j\}$, where the $\sigma_k > 0$ are scaling factors. The QR factorization of $N \times (m+1)$ matrices with columns $\sigma_k A^k r_j$, $0 \leq k \leq m$, is computed by using Householder transformations. Computed examples show this approach to yield poor or even no convergence. This is due to the fact that the bases $\{\sigma_k A^k r_j\}_{k=0}^m$ can be severely ill-conditioned. In the scheme in [3, 4] and the present paper, the scaled monomial bases are replaced by better conditioned bases of Newton form. This change of bases eliminates the numerical difficulties encountered with the monomial bases; see the computed examples in [3, 4].

We introduce notation necessary to describe our implementation. Let Π_{m-1} denote the set of polynomials of degree at most $m-1$. The problem (1.5) is then equivalent to the minimization problem

$$\min_{p\in\Pi_{m-1}} \|r_j - Ap(A)r_j\| \,, \tag{1.5'}$$

whose solution we denote by p_j. Then (1.6) can be written as

$$x_{j+1} := x_j + p_j(A)r_j \,. \tag{1.6'}$$

We represent $p \in \Pi_{m-1}$ by bases of Newton form $\left\{\sigma_k \prod_{l=1}^{k}(\cdot - \lambda_l)\right\}_{k=0}^{m-1}$, where the $\sigma_k > 0$ are scaling factors and the parameters $\lambda_l \in \mathbb{C}$ are chosen to make these bases fairly well-conditioned. Theoretical and computational results in [14] show that a polynomial basis of Newton form, with a suitable choice of parameters λ_l, can be much better conditioned than a basis of power form. In particular, the Newton basis is found to be fairly well-conditioned on a compact set S in the complex plane $\mathbb{C}$ when the λ_l are chosen to be Leja points (defined below) for S. This suggests that a good choice of parameters λ_l would be Leja points for $\lambda(A)$, the spectrum of A; a more detailed discussion on this choice of λ_l can be found in [3]. However, $\lambda(A)$ is not explicitly known. We therefore choose the λ_l to be Leja points for $\lambda(H_m)$, where the upper Hessenberg matrix $H_m = V_m^T A V_m$ is made up of the first m rows of the matrix H.

Let S be a compact set in $\mathbb{C}$. A set of points $\{\zeta_j\}_{j=1}^m \in$ S is said to be a set of *Leja points* for S if

$$|\zeta_1| = \max_{\zeta\in S} |\zeta|, \tag{1.7a}$$

$$\prod_{l=1}^{j} |\zeta_{j+1} - \zeta_l| = \max_{\zeta\in S} \prod_{l=1}^{j} |\zeta - \zeta_l|, \quad j = 1, 2, \ldots, m-1\,. \tag{1.7b}$$

Note that a set of m Leja points for a given set S might not be unique. Asymptotic properties of Leja points for compact sets of infinite cardinality were first studied by Edrei [8] and Leja [12].* In particular, if S is an interval, then the Leja points for S are distributed roughly like zeros of Chebyshev polynomials for S, and if instead S is a disk, then the Leja points for S are uniformly distributed on the boundary of S; see [8, 12, 14, 15].

Introduce the $N \times m$ matrix

$$B_m := \left[\sigma_0 r_j, \sigma_1(A - \lambda_1 I)r_j, \ldots, \sigma_{m-1} \prod_{l=1}^{m-1} (A - \lambda_l I)r_j\right]. \tag{1.8}$$

Then for any $p \in \Pi_{m-1}$ we have $p(A)r_j = B_m d$ for some vector $d \in \mathbb{C}^m$. The minimization problem (1.5') is equivalent to the problem

$$\min_{d\in\mathbb{C}^m} \|r_j - AB_m d\| \,, \tag{1.5''}$$

whose solution we denote by d_j. Then (1.6') becomes

$$x_{j+1} := x_j + B_m d_j \,. \tag{1.6''}$$

*Despite that the paper [8] precedes [12], points that satisfy (1.7b) are commonly referred to as Leja points.

Details of the algorithm for solving (1.5") are presented in Section 2. Section 3 presents our parallel implementation, and Section 4 describes computed examples.

Several other approaches, to obtain an implementation of Algorithm 1.2, that are suitable for parallel and vector computers are described in the literature; see, e.g., [5, 9, 10, 13, 18, 20, 21]. These approaches have in common that they seek to avoid the vector-vector operations of the MGS method in Algorithm 1.1. Similarly with the scheme in [3, 4] and the present paper, the algorithms by de Sturler [18] and Joubert and Carey [10] first generate certain nonorthogonal bases of Krylov subspaces $\mathrm{K}_{m+1}(A, r_j)$. These bases are then orthogonalized. The algorithms in [10, 18] can therefore also be regarded as generalizations of the approach in [9]. The approaches differ in the choice of nonorthogonal bases. Moreover, in [10] the least-squares problem (1.5) is solved by forming normal equations, rather than using a QR factorization. The ill-conditioning of the normal equations may lead to loss of accuracy or a reduced rate of convergence. Chronopoulos and Kim [5] implement an s-step version of the GMRES algorithm on a Cray-2 computer using one and four processors. Walker [20, 21] presents two schemes that use Householder transformations for computing orthonormal bases of Krylov subspaces $\mathrm{K}_{m+1}(A, r_j)$. However, the flop counts of the schemes in [20, 21] are approximately 2 and 3 times larger than for Algorithm 1.1, and this can make them slower than Algorithm 1.1. Pernice [13] reports the performance of the schemes by Walker [20] and Chronopoulos and Kim [5], when implemented on an IBM 3090 computer using one processor. It is presently difficult to compare these different approaches, since they differ in flop counts and communication requirements, and have not been implemented on the same computer.

2 A Newton basis GMRES algorithm

This section describes an implementation of Algorithm 1.2 that avoids the vector-vector operations in Algorithm 1.1. Our presentation follows [3, 4]. We first note that the GMRES method does not require the matrix A to be stored. It suffices to compute matrix-vector products Au for certain vectors $u \in \mathbb{R}^N$. In many applications the vector $\delta(A - \lambda I)u$, where δ and λ are real constants, can be evaluated almost as rapidly as the vector Au. Matrices A with this property arise from the discretization of partial differential and integral equations. This is illustrated by the matrices used in the computed examples of Section 4. It is convenient to introduce a vector-valued function $\mathrm{mult}(\delta, A, \lambda, u)$, whose value is $\delta(A - \lambda I)u$. This function can be evaluated conveniently in parallel.

Let x_0 be an initial approximate solution, $r_0 := b - Ax_0$ and m be the dimension of the Krylov subspace $\mathrm{K}_m(A, r_0)$. Then Algorithm 1.1 yields a new improved approximate solution x_1 and the $m \times m$ upper Hessenberg matrix H_m. The spectrum $\lambda(H_m) = \{\lambda_j\}_{j=1}^m$ can be computed by standard numerical software, e.g., by EISPACK [17] subroutine HQR.

Because A, b and x_0 contain real entries only, so does the matrix H_m. The eigenvalues λ_j of H_m are therefore real or appear in complex conjugate pairs. First assume that $\lambda(H_m)$ is real. We can then use formulas (1.7a) and (1.7b) with $S := \lambda(H_m)$ to order the eigenvalues of H_m. If the spectrum $\lambda(H_m)$ is not real, then we modify the ordering scheme (1.7a) and (1.7b) so that for the ordered eigenvalues λ_j we have that $\mathrm{Im}(\lambda_j) > 0$ implies $\lambda_{j+1} = \bar{\lambda}_j$, where the bar denotes complex conjugation; see [4, Algorithm 3.1] for details. This modification of (1.7a) and (1.7b) allows us to work with real vectors only.

We wish to determine the QR factorization of the matrix AB_m in (1.5"). To this end we first compute the matrix B_{m+1} given by (1.8) and then determine its QR factorization. The computation of the matrix B_{m+1} is described by the following algorithm, which determines a matrix $\tilde{B}_{m+1} = [\tilde{b}_1, \tilde{b}_2, \ldots, \tilde{b}_{m+1}] \in \mathbb{R}^{N\times(m+1)}$ and a diagonal matrix $\tilde{D}_{m+1} = \mathrm{diag}[\tilde{\delta}_1, \tilde{\delta}_2, \ldots, \tilde{\delta}_{m+1}]$, such that $B_{m+1} := \tilde{B}_{m+1}\tilde{D}_{m+1}$.

Algorithm 2.1 (Computation of $\tilde{B}_{m+1}$ and $\tilde{D}_{m+1}$)
Input: m Leja ordered eigenvalues $\lambda_1, \lambda_2, \ldots, \lambda_m$ *of* H_m*, residual* $r_1 := b - Ax_1$;
Output: matrices $\tilde{B}_{m+1}$, $\tilde{D}_{m+1}$;
$\tilde{b}_1 := r_1; \quad \tilde{\delta}_1 := 1/\|\tilde{b}_1\|;$
for $j := 1, 2, \ldots, m$ *do*
  if $\mathrm{Im}(\lambda_j) = 0$ *then*
    $\tilde{b}_{j+1} := \mathrm{mult}(\tilde{\delta}_j, A, \lambda_j, \tilde{b}_j); \quad \tilde{\delta}_{j+1} := 1/\|\tilde{b}_{j+1}\|;$
  else
    if $\mathrm{Im}(\lambda) > 0$ *then*

$$\begin{aligned} \tilde{b}_{j+1} &:= \mathrm{mult}\left(\tilde{\delta}_j, A, \mathrm{Re}(\lambda_j), \tilde{b}_j\right); \\ \tilde{b}_{j+2} &:= \mathrm{mult}\left(1, A, \mathrm{Re}(\lambda_j), \tilde{b}_{j+1}\right) + \tilde{\delta}_j \mathrm{Im}(\lambda_j)^2 \tilde{b}_j; \\ \tilde{\delta}_{j+1} &:= 1/\|\tilde{b}_{j+1}\|; \ \tilde{\delta}_{j+2} := 1/\|\tilde{b}_{j+2}\|; \end{aligned}$$

    end if
  end if
end j;

We compute the QR factorization

$$\tilde{B}_{m+1} = \tilde{Q}_{m+1}\tilde{R}_{m+1}, \tag{2.1}$$

where $\tilde{Q}_{m+1}$ is orthogonal and $\tilde{R}_{m+1}$ is upper triangular. This factorization can be computed without column pivoting.

The next step is to determine the QR factorization of the matrix AB_m, where B_m is given by (1.8). This can be accomplished with fairly little work, given the factorization (2.1). We use the fact that the columns $\tilde{b}_j$ of $\tilde{B}_{m+1}$ satisfy a recursion relation. If $\lambda_j \in \mathbb{R}$ then

$$\tilde{b}_{j+1} = \tilde{\delta}_j (A - \lambda_j I)\tilde{b}_j, \tag{2.2a}$$

and if $\mathrm{Im}(\lambda_j) > 0$, then

$$\tilde{b}_{j+1} = \tilde{\delta}_j \left(A - \mathrm{Re}(\lambda_j) I\right) \tilde{b}_j, \qquad \tilde{b}_{j+2} = \tilde{\delta}_j (A - \lambda_j I)(A - \bar{\lambda}_j I) b_j. \tag{2.2b}$$

Let $\tilde{Q}_{m+1} = [\tilde{q}_1, \tilde{q}_2, \ldots, \tilde{q}_{m+1}]$ and $\tilde{R}_{m+1} = [\tilde{r}_1, \tilde{r}_2, \ldots, \tilde{r}_{m+1}] = [\tilde{\rho}_{kj}]$. Then by (2.1),

$$\tilde{B}_{m+1}e_j = \tilde{b}_j = \tilde{Q}_{m+1}\tilde{r}_j = \sum_{k=1}^{j} \tilde{q}_k\tilde{\rho}_{kj}, \quad \text{for } 1 \leq j \leq m+1. \tag{2.3}$$

First assume that $\lambda_j \in \mathbb{R}$. Then, by (2.2a) and (2.3),

$$\begin{aligned} AB_{m+1}e_j &= A\tilde{B}_{m+1}\tilde{D}_{m+1}e_j = \tilde{\delta}_j A\tilde{b}_j = \tilde{b}_{j+1} + \tilde{\delta}_j\lambda_j\tilde{b}_j \\ &= \sum_{k=1}^{j} \tilde{q}_k(\tilde{\rho}_{k,j+1} + \tilde{\delta}_j\lambda_j\tilde{\rho}_{kj}) + \tilde{q}_{j+1}\tilde{\rho}_{j+1,j+1} = \sum_{k=1}^{j+1} \tilde{q}_k\hat{\rho}_{kj}, \end{aligned} \tag{2.4}$$

where

$$\begin{aligned} \hat{\rho}_{kj} &:= \tilde{\rho}_{k,j+1} + \tilde{\delta}_j\lambda_j\tilde{\rho}_{kj}, \quad 1 \leq k \leq j, \\ \hat{\rho}_{j+1,j} &:= \tilde{\rho}_{j+1,j+1}. \end{aligned} \tag{2.5}$$

If instead $\mathrm{Im}(\lambda_j) > 0$, then we obtain from (2.2b) and (2.3) that

$$AB_{m+1}e_j = \sum_{k=1}^{j+1} \tilde{q}_k\hat{\rho}_{kj}, \quad AB_{m+1}e_{j+1} = \sum_{k=1}^{j+2} \tilde{q}_k\hat{\rho}_{k,j+1}, \tag{2.6}$$

where

$$\begin{aligned} \hat{\rho}_{kj} &:= \tilde{\rho}_{k,j+1} + \tilde{\delta}_j\mathrm{Re}(\lambda_j)\tilde{\rho}_{kj}, \quad 1 \leq k \leq j, \\ \hat{\rho}_{j+1,j} &:= \tilde{\rho}_{j+1,j+1}, \\ \hat{\rho}_{k,j+1} &:= \tilde{\delta}_{j+1}\left(\tilde{\rho}_{k,j+2} + \mathrm{Re}(\lambda_j)\tilde{\rho}_{k,j+1} - \tilde{\delta}_j\mathrm{Im}(\lambda_j)^2\tilde{\rho}_{kj}\right), \quad 1 \leq k \leq j, \\ \hat{\rho}_{j+1,j+1} &:= \tilde{\delta}_{j+1}\left(\tilde{\rho}_{j+1,j+2} + \mathrm{Re}(\lambda_j)\tilde{\rho}_{j+1,j+1}\right), \\ \hat{\rho}_{j+2,j+1} &:= \tilde{\delta}_{j+1}\tilde{\rho}_{j+2,j+2}. \end{aligned} \tag{2.7}$$

It follows from (2.4) and (2.6) that the $(m+1) \times m$ matrix $\hat{R} = [\hat{\rho}_{jk}]$ satisfies

$$AB_m = \tilde{Q}_{m+1}\hat{R}, \tag{2.8}$$

where $\hat{\rho}_{jk} := 0$, if $1 < j+1 < k \leq m+1$. A QR factorization of $\hat{R}$ can be determined using only m Givens rotations. This factorization and (2.8) are used to solve (1.5"). We summarize our implementation as follows.

Algorithm 2.2 (A Newton basis GMRES algorithm)
Choose initial vector x_0, $\varepsilon > 0$ *and the dimension* m *of the Krylov subspace;*
$r_0 := b - Ax_0$;
Apply Algorithm 1.1 to determine x_1 *and* H_m;
$r_1 := b - Ax_1$;
Compute the spectrum $\lambda(H_m) = \{\lambda_j\}_{j=1}^m$ *and Leja order the* λ_j;
for $j := 1, 2, \ldots$ *until* $\|r_j\| \leq \varepsilon$ *do*
  Compute columns of $\tilde{B}_{m+1}$ *by Algorithm 2.1;*
  Compute QR factorization of $\tilde{B}_{m+1} = \tilde{Q}_{m+1}\tilde{R}_{m+1}$;
  Compute the $(m+1) \times m$ *matrix* $\hat{R}$ *by (2.5) and (2.7), so that* $AB_m = \tilde{Q}_{m+1}\hat{R}$;
  Compute QR factorization of $\hat{R}$, *solve (1.5") for* d_j;
  Compute $x_{j+1} := x_j + \tilde{B}_m\tilde{D}_m d_j$ *and* $r_{j+1} := b - Ax_{j+1}$;
end j;

We describes our parallel implementation of Algorithm 2.2 in Section 3. We conclude this section with two remarks on alternative ways to carry out the computations in the algorithm.

Remark 2.1. We can evaluate the residual vector according to $r_{j+1} := r_j - Q_{m+1}\hat{R}d_j$. This can be faster than using the formula $r_{j+1} := b - Ax_{j+1}$, but may yield lower accuracy and has therefore not been used.

Remark 2.2. Evaluation of

$$x_{j+1} := x_j + \tilde{B}_m\tilde{D}_m d_j \tag{2.9}$$

requires that the matrix $\tilde{B}_m$ be stored. Storage space can be saved by instead evaluating x_{j+1} according to

$$x_{j+1} := x_j + Q_m R_m d_j, \tag{2.10}$$

where Q_m consists of the first m columns of Q_{m+1} and R_m is the leading principal $m \times m$ submatrix of R_{m+1}. If (2.10) is used, then Q_{m+1} can overwrite B_{m+1}. However, since many subroutines for QR factorization store the matrix Q_m in factored form, e.g., the subroutine that we use in our implementation stores Q_m as a product of Householder transformations, the evaluation of x_{j+1} by (2.10) can be slower than when (2.9) is used. We evaluate x_{j+1} by (2.9) in our implementation since plenty of storage space is available on the IBM 3090 computer used for our timing experiments.

3 A parallel implementation

We describe the parallel implementation used in the computed examples. The computation of the QR factorization of the matrix $\tilde{B}_{m+1}$ is carried out in parallel, and so is the evaluation of vectors $\delta(A - \lambda I)u$ (the function mult).

The time measurements presented in this paper are for a parallel FORTRAN implementation of Algorithm 2.2 using up to four vector processors. Our implementation utilizes level 1 and level 2 BLAS. Numerical experiments in [3, 4] show that the sequential version of Algorithm 2.2 which utilizes level 1 and level 2 BLAS requires 10-12% less CPU time than a GMRES implementation based on Algorithm 1.1 that uses level 1 BLAS. Furthermore, it appeared that 85% of the total CPU time required is spent computing the QR factorization of the matrices $\tilde{B}_{m+1}$.

In our parallel implementation of Algorithm 2.2 we compute the QR factorization of the matrix $\tilde{B}_{m+1}$ by using a simplified version of a scheme described by Chu and George in [6] for the QR factorization of a dense matrix on shared memory multiprocessors.

We describe our implementation for the case when four processors are used, and comment on the case when a different number of processors are used in the end of this section. First we partition the matrix $\tilde{B}_{m+1}$ into four disjoint blocks of $N/4$ contiguous rows. Each independent processor is assigned to compute the QR factorization of one block. The LAPACK subroutine XGEQR2 [1], which utilizes both level 1 and level 2 BLAS, is used to determine the QR factorization of each block. In this part of the computations there are essentially no synchronization cost or idle time because each processor is assigned the same amount of work. The parallel reduction of each block to upper triangular form reduces $\tilde{B}_{m+1}$ to a matrix of the form

$$B'_{m+1} = \begin{bmatrix} \times & \times & \times \\ & \times & \times \\ & & \times \\ \vdots & \vdots & \vdots \\ \times & \times & \times \\ & \times & \times \\ & & \times \\ \vdots & \vdots & \vdots \\ \times & \times & \times \\ & \times & \times \\ & & \times \\ \vdots & \vdots & \vdots \\ \times & \times & \times \\ & \times & \times \\ & & \times \\ \vdots & \vdots & \vdots \end{bmatrix}, \tag{3.1}$$

where the symbol $\times$ denotes an entry that may be nonvanishing. The matrix shown in (3.1) corresponds to $m = 2$. Having computed B'_{m+1}, we define the $2(m+1) \times (m+1)$ matrix $B''_{m+1,1}$ as the union of the top two $(m+1) \times (m+1)$ nonvanishing upper triangular submatrices of B'_{m+1}. Similarly, we define the $2(m+1) \times (m+1)$ matrix $B''_{m+1,2}$ as the union of the bottom two $(m+1) \times (m+1)$ nonvanishing upper triangular submatrices of B'_{m+1}. The matrices $B''_{m+1,j}$ are of the form

$$B''_{m+1,j} = \begin{bmatrix} \times & \times & \times \\ & \times & \times \\ & & \times \\ \times & \times & \times \\ & \times & \times \\ & & \times \end{bmatrix}. \tag{3.2}$$

The QR factorization of the two matrices $B''_{m+1,j}$ is executed in parallel by two independent processors using the subroutine XGEQR2. This yields a new pair of triangular matrices, which are merged to yield the $2(m+1) \times (m+1)$ matrix B'''_{m+1} of the form displayed in (3.2). The QR factorization of B'''_{m+1} by subroutine XGEQR2 yields the upper triangular matrix $\tilde{R}_{m+1}$. Note that each time two

$(m+1) \times (m+1)$ submatrices are merged into a $2(m+1) \times (m+1)$ matrix, the number of processors used concurrently is reduced, and, thus, the idle time is increased. However, due to the fact that in applications of the GMRES method generally $N/4 \gg m+1$, the by far largest amount of computing time is spent on the QR factorization of the first four blocks.

It is now obvious how to modify our implementation when we have $\kappa \neq 4$ processors available. We first partition the matrix $\tilde{B}_{m+1}$ into κ disjoint blocks of N/κ contiguous rows. Each independent processor is assigned to compute the QR factorization of one block. Under the assumption that $N/\kappa \gg m+1$, most of the arithmetic work required for one iteration is spent on the QR factorization of these blocks. Having determined κ triangular factors in parallel, we merge pairs of these factors into $2(m+1) \times (m+1)$ matrices of the form (3.2). The QR factorization of these matrices can be computed in parallel. This yields new triangular factors, which we merge into larger matrices, whose QR factorization we determine. We proceed in this manner until the triangular $(m+1) \times (m+1)$ matrix in the QR factorization of $\tilde{B}_{m+1}$ has determined. Details of the scheme depend on the value of κ. The algorithm is most simple to implement when κ is a power of two.

4 Numerical experiments

Our numerical experiments were performed on an IBM 3090-600S VF computer. An overview of the machine architecture is presented in [19]. All codes were written in double precision arithmetic (about 15 significant digits) and IBM's VS FORTRAN Version 2 Release 5, which contains a number of parallel extensions to standard FORTRAN. The parallel extensions used in our code are Parallel Do and Parallel Call. Parallel Do splits a do-loop across several processors and Parallel Call allows subroutines to be executed simultaneously.

We derived our test problem by discretizing the boundary value problem

$$-\Delta u + 2p_1 u_x + 2p_2 u_y - p_3 u = f \quad \text{in} \quad \Omega \tag{4.1}$$

with $u = 0$ on $\partial\Omega$ by finite differences, where Ω is the unit square $\{(x,y) \in R^2, 0 < x, y < 1\}$ and p_1, p_2, p_3 are positive constants. The function $f(x,y)$ is chosen so that $u(x,y) = xe^{xy}\sin(\pi x)\sin(\pi y)$ solves (4.1). More precisely, (4.1) is discretized by centered finite differences on a uniform $n \times n$ grid, using the standard five-point stencil to approximate the Laplacian. Introduce $h := 1/(n+1)$. After scaling by h^2, the algebraic linear system of equations of order $n^2 \times n^2$ obtained from (4.1) can be written as (1.1) with $N = n^2$, where a typical equation for the unknown $u_{ij} \approx u(ih, jh)$ is given by

$$(4-\sigma)u_{ij} - (1+\beta)u_{i-1,j} - (1-\beta)u_{i+1,j} - (1+\gamma)u_{i,j-1} - (1-\gamma)u_{i,j+1} = h^2 f_{ij}.$$

Here $\beta := p_1 h, \gamma := p_2 h, \sigma := p_3 h^2$ and $f_{ij} := f(ih, jh)$. No preconditioner was used in order to keep the issues of interest clear. Generally, however, a preconditioner should be used; see [2, 7, 16, 20] for discussions and illustrations.

$N\backslash m$	10	30	50
100^2	.15	.74	1.75
120^2	.22	1.05	2.48
140^2	.30	1.43	3.31
160^2	.38	1.84	4.25
180^2	.48	2.28	5.43
200^2	.60	2.90	7.01

Table 1: Timing for one cycle of the j-loop in Algorithm 2.2 with one processor

$N\backslash m$	10	30	50
100^2	.11	.47	1.05
120^2	.15	.62	1.47
140^2	.20	.85	1.92
160^2	.24	1.13	2.44
180^2	.30	1.37	3.12
200^2	.38	1.75	4.13

Table 2: Timing for one cycle of the j-loop in Algorithm 2.2 with two processors

Time measurements were done by recording the wall clock time on a fully dedicated machine. The calculations of execution times per cycle, displayed in Tables 1-3, were obtained by subtracting the execution time for Algorithm 2.2, with zero cycles in the j-loop, from the execution time with 20 cycles in the j-loop, and then dividing the remaining time by 20.

Table 1 presents timings for the parallel code utilizing only one processor. The square root of the number of rows of the matrix $\tilde{B}_{m+1}$ goes from 100 to 200 through increments of 20. The size of the Krylov subspace m is equal to 10, 30, and 50. The execution times in Table 1 are very close to the execution times for the sequential version of the code in which the matrix $\tilde{B}_{m+1}$ is not partitioned into blocks.

$N\backslash m$	10	30	50
100^2	.087	.33	.71
120^2	.12	.46	1.03
140^2	.15	.60	1.36
160^2	.18	.78	1.79
180^2	.25	.99	2.29
200^2	.30	1.31	2.97

Table 3: Timing for one cycle of the j-loop in Algorithm 2.2 with four processors

$N\backslash m$	10	30	50
100^2	1.36	1.57	1.67
120^2	1.47	1.69	1.69
140^2	1.50	1.68	1.72
160^2	1.58	1.63	1.74
180^2	1.60	1.66	1.74
200^2	1.58	1.66	1.70

Table 4: Speedup with two processors

$N\backslash m$	10	30	50
100^2	1.72	2.24	2.46
120^2	1.83	2.28	2.24
140^2	2.00	2.38	2.43
160^2	2.11	2.36	2.37
180^2	1.92	2.30	2.37
200^2	2.00	2.21	2.36

Table 5: Speedup with four processors

Tables 2-3 display the execution times for one cycle of the j-loop in Algorithm 2.2 using two and four processors. Tables 4-5 display the speedup achieved with two and four processors. The speedup is computed according to the formula

$$speedup := \frac{t(N,m,1)}{t(N,m,p)},$$

where $t(N,m,q)$ is the execution time for one cycle of the j-loop of Algorithm 2.2 using q processors. The speedup for two processors increases with the size of the matrix up to a maximum value of 1.74 for $N = 160^2$ or 180^2 and $m = 50$. The measured speedups are compatible with a parallel part of over 85%. In view of that 85% of the total CPU time for the execution of the sequential code was spent on the QR factorization of the matrices $\tilde{B}_{m+1}$, the measured speedup for two processors is quite satisfactory.

The speedup for four processors also increases with the size of the matrix, reaching a maximum value of 2.46 when $N = 100^2$ and $m = 50$. The speedup for $m \geq 30$ is compatible with a parallel part of 55% to 60%. We recall that in the QR factorization of the matrices $\tilde{B}_{m+1}$ four processors are used only in the first phase of the computations, and this explains, in part, why the speedup for 4 processors is not twice the speedup for 2 processors.

$N \backslash m$	10	30	50
100^2	68%	79%	83%
120^2	73%	85%	84%
140^2	75%	84%	86%
160^2	79%	81%	87%
180^2	80%	83%	87%
200^2	79%	83%	85%

Table 6: Efficiency with two processors

$N \backslash m$	10	30	50
100^2	43%	56%	62%
120^2	46%	57%	60%
140^2	50%	60%	61%
160^2	53%	59%	59%
180^2	48%	57%	59%
200^2	50%	56%	59%

Table 7: Efficiency with four processors

The efficiency $E(N, m, q)$ is defined as the ratio of the wall-clock time for the completion of one cycle of the j-loop in Algorithm 2.2 utilizing only one processor, to the corresponding time using q processors multiplied by the number of processors. Tables 6-7 show the efficiency of our implementation of the GMRES algorithm for 2 and 4 processors. The efficiency for two processors ranges from 68% for the smallest problem ($N = 100^2$, $m = 10$) to 87% when $N = 160^2$ and $m = 50$. Note that, even though speedup and efficiency increase with N for each fixed value of m, when we compare speedup and efficiency for matrices $\tilde{B}_{m+1}$ with the same number of elements (i.e., $N \times (m + 1)$ is kept fixed), we find that the higher efficiency and speedup are observed for larger values of m.

5 Conclusions

The paper describes a parallel implementation of the Newton basis GMRES algorithm for solving large sparse linear systems. In this implementation the QR factorization, of dense matrices $\tilde{B}_{m+1}$ with many rows, is assigned to four independent processors according to a simplified version of a scheme for QR factorization on shared memory multiprocessors proposed by Chu and George [6].

The numerical experiments were carried out on an IBM 3090-600S VF computer. Execution times for one cycle of the j-loop in Algorithm 2.2 employing 1,

2 or 4 processors are presented. Improvements in speed of up to a factor of 1.74 are observed for larger problems when using 2 processors, and up to a factor of 2.46 when using 4 processors. The efficiency ranges from 80% to 87% for $m \geq 30$ with two processors, and form 56% to 61% for $m \geq 30$ with four processors.

Acknowledgements. The authors would like to thank Ronny Andersen for help with the timings and Stratis Gallopoulos and Richard Varga for comments on a previous version of the manuscript. D.C. and L.R. would like to thank Pat Gaffney for an enjoyable stay at BSC.

References

[1] E. ANDERSON, Z. BAI, C. BISCHOF, J. DEMMEL, J. DONGARRA, J. DU CROZ, A. GREENBAUM, S. HAMMARLING, A. MCKENNEY, S. OSTROUCHOV AND D. SORENSEN, *LAPACK Users' Guide*, SIAM, Philadelphia, 1992.

[2] E. ANDERSON AND Y. SAAD, *Preconditioned conjugate gradient methods for general sparse matrices on shared memory computers,* in Parallel Processing for Scientific Computing, ed. G. Rodrigue, SIAM, Philadelphia, 1989, pp. 88–92.

[3] Z. BAI, D. HU AND L. REICHEL, *An implementation of the GMRES method using QR factorization*, in Proceedings of the Fifth SIAM Conference on Parallel Processing for Scientific Computing, eds. J. Dongarra, K. Kennedy, P. Messina, D.C. Sorensen and R.G. Voigt, SIAM, Philadelphia, 1992, to appear.

[4] Z. BAI, D. HU AND L. REICHEL, *A Newton basis GMRES implementation*, Report 91-03, Department of Mathematics, University of Kentucky, April 1991 (to appear in IMA J. Numer. Anal.).

[5] A.T. CHRONOPOULOS AND S.K. KIM, *s-step orthomin and GMRES implemented on parallel computers*, Report TR 90-15, Computer Science Dept., University of Minnesota, Twin Cities, 1990.

[6] E. CHU AND A. GEORGE, *QR factorization of a dense matrix on a shared-memory multiprocessor,* Parallel Comput., 11 (1989), pp. 55–71.

[7] J.J. DONGARRA, I.S. DUFF, D.C. SORENSEN AND H.A. VAN DER VORST, *Linear System Solving on Vector and Shared Memory Computers,* SIAM, Philadelphia, 1991.

[8] A. EDREI, *Sur les déterminants récurrents et les singularités d'une fonction donneé par con développement de Taylor,* Composito Math., 7 (1939), pp. 20–88.

[9] A.C. HINDMARSH AND H.F. WALKER, *Note on a Householder implementation of the GMRES method,* Report UCID-20899, Lawrence Livermore National Laboratory, 1986.

[10] W.D. Joubert and G.F. Carey, *Parallelizable restarted iterative methods for nonsymmetric linear systems. Part I: Theory, Part II: Parallel Implementation,* Report CNA-251, Center for Numerical Analysis, the University of Texas at Austin, May 1991, submitted to International Journal of Computer Mathematics, Gordon Breach publ., special issue on preconditioned conjugate gradient methods.

[11] R. Karlsson, *A study of some roundoff error effects of the GMRES method,* Report LiTH-Math-R-1990-11, Department of Mathematics, Linköping University, Linköping, Sweden, 1991.

[12] F. Leja, *Sur certaines suites liées aux ensemble plans et leur application à la representation conforme,* Ann. Polon. Math., 4 (1957), pp. 8–13.

[13] M. Pernice, *Implementations of GMRES and their performance on the IBM 3090/600S,* USI Report No. 13, Utah Supercomputing Institute, University of Utah, Salt Lake City, 1991.

[14] L. Reichel, *Newton interpolation at Leja points,* BIT, 30 (1990), pp. 332–346.

[15] L. Reichel, *The application of Leja points to Richardson iteration and polynomial preconditioning,* Linear Algebra Appl., 154/156 (1991), pp. 389–414.

[16] Y. Saad and M.H. Schultz, *GMRES: A generalized minimal residual algorithm for solving nonsymmetric linear systems,* SIAM J. Sci. Statist. Comput., 7 (1986), pp. 856–869.

[17] B.T. Smith, J.M. Boyle, Y. Ikebe, V.C. Klema, and C.B. Moler, *Matrix Eigensystem Routines: EISPACK Guide,* 2nd ed., Springer, New York, 1970.

[18] E. de Sturler, *A parallel variant of GMRES(m),* in Proceedings of the 13th World Congress on Computation and Applied Mathematics, Dublin, Ireland (IMACS '91), International Association for Mathematics and Computers in Simulation, 1991, pp. 682–683.

[19] S.G. Tucker, *The IBM 3090 system: an overview,* IBM System J., 25 (1986), pp. 4–19.

[20] H.F. Walker, *Implementation of the GMRES method using Householder transformations,* SIAM J. Sci. Statist. Comput., 9 (1988), pp. 152–163.

[21] H.F. Walker, *Implementations of the GMRES method,* Comput. Phys. Comm., 53 (1989), pp. 311–320.

Optimal semi-iterative methods applied to SOR in the mixed case

Michael Eiermann [*] *and Richard S. Varga* [†‡]

Abstract. The application of optimal semi-iterative methods to the standard successive over-relaxation (SOR) iterative method, with any real relaxation parameter ω, is completely analyzed here, under the assumptions that the associated Jacobi matrix B is consistently ordered and weakly cyclic of index 2 and that the spectrum, $\sigma(B^2)$, of B^2 satisfies $\sigma(B^2) \subset [-\alpha^2, \beta^2]$ with $0 < \alpha < \infty$ and $0 < \beta < 1$. The spectrum of B^2 is then a mixture of positive and negative eigenvalues, the so-called "mixed case". If $\kappa(\Omega_{\omega,\alpha,\beta})$ denotes the optimal asymptotic convergence factor for semi-iteration applied to $\mathcal{L}_\omega$ (the associated SOR iteration matrix), we deduce that

$$1 > \min_{\omega \in \mathbb{R}} \rho(\mathcal{L}_\omega) > \min_{\omega \in \mathbb{R}} \kappa(\Omega_{\omega,\alpha,\beta}) = [(\sqrt{1+\alpha^2} - \sqrt{1-\beta^2})^2]/[\alpha^2 + \beta^2] .$$

Key words. SOR, optimal semi-iterative methods, asymptotic convergence factor, covering domains.

AMS(MOS) subject classifications. 65F10.

1 Introduction

Recently in [5], the application of optimal semi-iteration to the standard successive overrelaxation (SOR) iterative method, with any real relaxation factor ω, was analyzed under the assumptions that the associated Jacobi matrix B is consistently ordered and weakly cyclic of index 2 and that the spectrum, $\sigma(B^2)$, of B^2 satisfies $\sigma(B^2) \subset [0, \beta^2]$, where $0 < \beta = \rho(B) < 1$. (This is the so-called "nonnegative case".) It was shown in [5, Theorem 1] that *no* semi-iterative method applied to the SOR method (for any real relaxation parameter ω) is asymptotically faster than the SOR method with optimal relaxation parameter $\omega = \omega_b$. The same statement is valid in the "nonpositive case" where $\sigma(B^2) \subset [-\alpha^2, 0]$ with $0 < \alpha = \rho(B)$ (cf. Theorem 1 in Section 2.

Here, we extend the results of [5], using similar techniques, to analyze the case where the Jacobi matrix B is consistently ordered and weakly cyclic of index 2

[*]Institut für Praktische Mathematik, Universität Karlsruhe, D-7500 Karlsruhe 1, F.R.G.
[†]Institute for Computational Mathematics, Kent State University, Kent, OH 44242, USA.
[‡]Research supported in part by the National Science Foundation.

Numerical Linear Algebra

with $\sigma(B^2) \subset [-\alpha^2, \beta^2]$ with $0 < \alpha < \infty$ and $0 < \beta < 1$ (the so-called "mixed case"). In contrast to the nonpositive and to the nonnegative case, there always exist semi-iterative methods which *improve* the asymptotic rate of convergence of the SOR method (with optimal relaxation parameter ω).

The outline of this paper is as follows: In Section 2, we review classical results on SOR and semi-iterative methods applied to SOR. The main theorem of this paper will be stated in Section 3 and proven in Section 5. The tools necessary for this proof will be developed in Section 4. Finally, in Section 6, we briefly comment on the question as to what extent our results carry over to the p-cyclic case.

2 A review of classical results on SOR

Consider the system of linear equations

$$A\mathbf{x} = \mathbf{b}, \text{ where } A \in \mathbb{R}^{N\times N} \text{ and } \mathbf{b} \in \mathbb{R}^N \text{ are given,} \tag{1}$$

with the standard splitting of the coefficient matrix A,

$$A = D - L - U,$$

where D is a nonsingular block diagonal matrix, and where L and U denote respectively strictly lower and strictly upper triangular matrices. We further assume that the corresponding block Jacobi matrix B, defined by

$$B := D^{-1}(L+U), \tag{2}$$

is consistently ordered and weakly cyclic of index 2 (cf. [9, Definition 4.2]), and that the spectrum, $\sigma(B^2)$, of the matrix B^2 consists of real numbers satisfying

$$\sigma(B^2) \subset [-\alpha^2, \beta^2] \text{ with } 0 \le \alpha < \infty \text{ and } 0 \le \beta < 1. \tag{3}$$

We further assume that the interval $[-\alpha^2, \beta^2]$ is a sharp enclosure of $\sigma(B^2)$, i.e., that

$$-\alpha^2, \beta^2 \in \sigma(B^2).$$

The assumption (3) implies that there is a unique solution $\mathbf{x}$ to the matrix equation (1). Since the case $\alpha = \beta = 0$ is essentially trivial and since the case of $\alpha = 0$ and $0 < \beta < 1$ was treated in [5, Theorem 1];* we assume that

$$\sigma(B^2) \subset [-\alpha^2, \beta^2] \text{ with } -\alpha^2, \beta^2 \in \sigma(B^2),\ 0 < \alpha < \infty \text{ and } 0 < \beta < 1. \tag{4}$$

We next review classical results for the SOR iterative method

$$\mathbf{x}_m = \mathcal{L}_\omega \mathbf{x}_{m-1} + \mathbf{c}_\omega \qquad (m = 1, 2, \ldots), \tag{5}$$

where $\mathcal{L}_\omega$, the SOR matrix, and $\mathbf{c}_\omega$ are defined by

$$\mathcal{L}_\omega := (D - \omega L)^{-1}[(1-\omega)D + \omega U], \text{ and } \mathbf{c}_\omega := \omega(D - \omega L)^{-1}\mathbf{b} \qquad (\omega \in \mathbb{R}). \tag{6}$$

*The case of $0 < \alpha < \infty$ and $\beta = 0$ can be analogously analyzed, and we shall briefly discuss this at the end of this section.

Here, ω is the associated *relaxation parameter.* Under the given assumptions on B and ω, the SOR method of (5) converges (for any initial vector $\mathbf{x}_0$) to the solution of (1) if and only if

$$0 < \omega < \frac{2}{1+\alpha} \tag{7}$$

holds (cf. Young [11, p. 193]). The optimal relaxation parameter ω_b which minimizes $\rho(\mathcal{L}_\omega)$, the spectral radius of $\mathcal{L}_\omega$, as a function of ω, is given by (cf. [11, p. 195])

$$\omega_b = \frac{2}{1+\sqrt{1+a^2-\beta^2}}, \tag{8}$$

and there also holds

$$1 > \rho(\mathcal{L}_\omega) > \rho(\mathcal{L}_{\omega_b}) = \left(\frac{\alpha+\beta}{1+\sqrt{1+\alpha^2-\beta^2}}\right)^2, \tag{9}$$

for all $0 < \omega < 2/(1+\alpha)$ with $\omega \neq \omega_b$.

As in [5], we now apply, for any fixed real ω, a semi-iterative method to the iterates $\{\mathbf{x}_m\}_{m=0}^{\infty}$ which are generated from the SOR iterations of (5), i.e., we consider vector sequences $\{\mathbf{y}_m\}_{m=0}^{\infty}$ of the form

$$\mathbf{y}_m := \sum_{j=0}^{m} \pi_{m,j}\mathbf{x}_j, \qquad (m = 0, 1, \ldots), \tag{10}$$

where the coefficients $\pi_{m,j}$ are (complex) constants which satisfy the constraint $\sum_{j=0}^{m} \pi_{m,j} = 1 \ \ (m = 0, 1, \ldots)$.

If $1 \notin \sigma(\mathcal{L}_\omega)$, it is well-known (cf. [9, p. 134]) that the associated error vectors $\mathbf{e}_m := (I - \mathcal{L}_\omega)^{-1}\mathbf{c}_\omega - \mathbf{y}_m$, for this semi-iterative method based on the basic iterative method of (5), satisfy

$$\mathbf{e}_m = p_m(\mathcal{L}_\omega)\mathbf{e}_0 \qquad (m = 0, 1, \ldots),$$

where $p_m(z) := \sum_{j=0}^{m} \pi_{m,j} z^j \in \Pi_m$, so that $p_m(1) = 1$. (Here, Π_m denotes the collection of all complex polynomials of degree at most m.)

For a given polynomial sequence $\{p_m\}_{m=0}^{\infty}$, with $p_m \in \Pi_m$ and $p_m(1) = 1$ for each $m \geq 0$, and for $1 \notin \sigma(\mathcal{L}_\omega)$, the quantity

$$\kappa\left(\mathcal{L}_\omega, \{p_m\}_{m=0}^{\infty}\right) := \limsup_{m\to\infty} \sup_{\mathbf{e}_0 \neq \mathbf{0}} \left[\frac{\|\mathbf{e}_m\|}{\|\mathbf{e}_0\|}\right]^{1/m}$$

(which depends on the structure of the Jordan canonical form of the matrix $\mathcal{L}_\omega$, but is independent of the vector norm $\|\cdot\|$ chosen on $\mathbf{R}^N$) measures the *asymptotic decay* of the norms of the error vectors $\mathbf{e}_m$ associated with (10).

As we shall see in Section 4 the eigenvalue assumption (4) leads to sharp inclusions $\sigma(\mathcal{L}_\omega) \subseteq \Omega = \Omega_{\omega,\alpha,\beta}$ for the spectrum of the SOR matrix $\mathcal{L}_\omega$, where $\Omega \subset \mathbf{C}$ is a compact set, and we call such a set Ω a *covering domain* for $\sigma(\mathcal{L}_\omega)$. In this setting, the best, i.e., *smallest*, asymptotic convergence factor for Ω which we can hope to achieve by *any* semi-iterative method is given by

$$\kappa(\Omega) := \lim_{m\to\infty} \left[\min\left\{\max_{z\in\Omega} |p(z)| \; : \; p \in \Pi_m, \; p(1) = 1\right\}\right]^{1/m}. \tag{11}$$

The quantity $\kappa(\Omega)$ is called the *asymptotic convergence factor* of the covering domain Ω, and this has been extensively studied in [4]. Note also that the definition of $\kappa(\Omega)$ in (11) couples *complex approximation theory* to the study of such semi-iterative methods.

With respect to the information $\sigma(\mathcal{L}_\omega) \subset \Omega$, the rate of convergence of the SOR iterative method (5) can therefore be improved, by the application of a semi-iterative method of the form (10), *only if*

$$\min\{1, \rho(\mathcal{L}_\omega)\} > \kappa(\Omega)$$

holds. The exact (nonempty) set of real ω's, for which the above inequality holds, will be precisely determined in Theorem 2 below.

How does one actually determine $\kappa(\Omega)$? We first note that $\kappa(\Omega) = 1$ holds for every compact set $\Omega \subset \mathbb{C}$ with $1 \in \Omega$. (For, if $1 \in \Omega$, then $\max_{z\in\Omega} |p(z)| \geq 1$ for any polynomial $p \in \Pi_m$ with $p(1) = 1$.) The same conclusion can also be drawn for another type of compact set Ω: Suppose that the complement $\mathbb{C}_\infty \backslash \Omega$ of Ω (with respect to the extended complex plane $\mathbb{C}_\infty$) is not connected, and that the points $z = 1$ and ∞ belong to different components of $\mathbb{C}_\infty \backslash \Omega$; then $\kappa(\Omega) = 1$. (In this case, there is a component of $\mathbb{C}_\infty \backslash \Omega$ containing $z = 1$, whose closure is compact and contains points of Ω. Then by the maximum principle, $\max_{z\in\Omega} |p(z)| \geq 1$ for any $p \in \Pi_m$ with $p(1) = 1$.)

Finally, on defining the class $\mathbb{M}$ by

$$\begin{aligned} \mathbb{M} := \quad & \{\Omega \subset \mathbb{C} : \Omega \text{ is compact and consists of more than one point,} \\ & \text{does not contain the point } z = 1, \\ & \text{and its complement } \mathbb{C}_\infty \backslash \Omega \text{ is simply connected}\}, \end{aligned} \tag{12}$$

then for $\Omega \in \mathbb{M}$,

$$\kappa(\Omega) = \frac{1}{|\Phi(1)|} \tag{13}$$

(cf. [4, Theorem 11]), where Φ is a conformal map from $\mathbb{C}_\infty \backslash \Omega$ onto the exterior of the unit circle with $\Phi(\infty) = \infty$. (We note, by the Riemann Mapping Theorem, that Φ exists and is unique, up to a constant factor of modulus 1.) Thus, if $\Omega \in \mathbb{M}$, the problem of determining its asymptotic convergence factor, $\kappa(\Omega)$, is reduced, from (13), to a problem in *conformal mapping theory.*

We mention that sharp covering domains Ω for the spectrum $\sigma(\mathcal{L}_\omega)$ will be determined and analyzed in Section4 for all real ω. It turns out, for fixed α and β with $0 < \alpha < \infty$ and $0 < \beta < 1$ and for any real ω, that there are essentially only *three* different types of covering domains $\Omega = \Omega_{\omega,\alpha,\beta}$ which need to be analyzed.

To conclude this section, we briefly describe the nonpositive case, where $\beta = 0$, i.e., B of (2) is consistently ordered and weakly cyclic of index 2 with

$$\sigma(B^2) \subset [-\alpha^2, 0] \text{ and } 0 < \alpha = \rho(B).$$

Matrices of this type arise for example in connection with a discretization of Theodorsen's integral equation (cf. Niethammer [8]). The optimal relaxation parameter ω_b, which minimizes $\rho(\mathcal{L}_\omega)$ as a function of ω, is then given by (cf. (8))

$$\omega_b = \frac{2}{1 + \sqrt{1 + \alpha^2}} \quad (< 1),$$

and (cf. (9))

$$\rho(\mathcal{L}_{\omega_b}) = 1 - \omega_b = \left(\frac{\alpha}{1+\sqrt{1+\alpha^2}}\right)^2 .$$

In the nonpositive case it turns out, similar to the nonnegative case, that *no* semi-iterative method applied to the SOR method (for any real relaxation parameter ω) is asymptotically faster than the SOR method with optimal relaxation parameter $\omega = \omega_b$ (see Figure 1).

Theorem 1. *Assume that the Jacobi matrix B of (2) is a consistently ordered weakly cyclic of index 2 matrix, and that the eigenvalues of B^2 are all nonpositive and lie in $[-\alpha^2, 0]$, where $0 < \alpha = \rho(B)$. Then, there holds*

$$\min_{\omega \in \mathbf{R}} \rho(\mathcal{L}_\omega) = \min_{\omega \in \mathbf{R}} \kappa(\Omega_{\omega,\alpha,0}) = 1 - \omega_b = \left(\frac{\alpha}{1+\sqrt{1+\alpha^2}}\right)^2 .$$

More precisely, the asymptotic convergence factor $\kappa(\Omega_{\omega,\alpha,0})$ has the following properties:

(i) *For $-\infty \leq \omega < \omega_1 := 2/(1-\sqrt{1+\alpha^2})$, $\kappa(\Omega_{\omega,\alpha,0}) = 1$, i.e., for these values of ω, no semi-iterative method applied to $\mathcal{L}_\omega$ converges.*

(ii) *For $\omega_1 < \omega < \omega_b$ and $\omega \neq 0$, $\kappa(\Omega_{\omega,\alpha,0})$ is a strictly decreasing function of ω which satisfies*

$$\min\{1, \rho(\mathcal{L}_\omega)\} > \kappa(\Omega_{\omega,\alpha,0}) > 1 - \omega_b .$$

(iii) *For $\omega_b \leq \omega \leq 1$, $\kappa(\Omega_{\omega,\alpha,0})$ is a constant function of ω which satisfies*

$$\min\{1, \rho(\mathcal{L}_\omega)\} \geq \kappa(\Omega_{\omega,\alpha,0}) = 1 - \omega_b$$

(where equality holds in the first inequality if and only if $\omega = \omega_b$),

(iv) *For $1 < \omega$, $\kappa(\Omega_{\omega,\alpha,0})$ is a strictly increasing function of ω which satisfies*

$$\min\{1, \rho(\mathcal{L}_\omega)\} > \kappa(\Omega_{\omega,\alpha,0}) > 1 - \omega_b .$$

3 Statement of the main result

To state our main result, Theorem 2 below, we define the following four specific real values of the relaxation factor ω (which are functions of α and β):

$$\omega_1 := \frac{2}{1-\sqrt{1+\alpha^2}}, \qquad \omega_2 := \frac{2}{1+\sqrt{1+\alpha^2}},$$
$$\omega_3 := \frac{2}{1+\sqrt{1-\beta^2}}, \qquad \omega_4 := \frac{2}{1-\sqrt{1-\beta^2}}. \tag{14}$$

We note that the assumptions $0 < \alpha < \infty$ and $0 < \beta < 1$ imply that

$$-\infty < \omega_1 < 0 < \omega_2 < 1 < \omega_3 < 2 < \omega_4 < \infty .$$

With the notation of (14), we come to the statement of our main result.

Theorem 2. *Assume that the Jacobi matrix B of (2) is a consistently ordered weakly cyclic of index 2 matrix, and that the eigenvalues of B^2 are (cf. (3)) all real and lie in $[-\alpha^2, \beta^2]$, where $0 < \alpha < \infty$, $0 < \beta < 1$ and $-\alpha^2, \beta^2 \in \sigma(B)$. Then, there holds*

$$\min\{1, \rho(\mathcal{L}_\omega)\} > \kappa(\Omega_{\omega,\alpha,\beta}) \quad (\omega \in \mathbb{R}) \tag{15}$$

if and only if $\omega \in (\omega_1, \omega_4) \setminus \{0\}$. More precisely, the asymptotic convergence factor $\kappa(\Omega_{\omega,\alpha,\beta})$ has the following properties:

(i) *For $\omega_1 < \omega < \omega_2$ and $\omega \neq 0$, $\kappa(\Omega_{\omega,\alpha,\beta})$ is a strictly decreasing function of ω which satisfies*

$$\min\{1, \rho(\mathcal{L}_\omega)\} > \kappa(\Omega_{\omega,\alpha,\beta}) > \frac{\left(\sqrt{1+\alpha^2} - \sqrt{1-\beta^2}\right)^2}{\alpha^2 + \beta^2}, \tag{16}$$

with $\lim_{\omega \uparrow 0} \kappa(\Omega_{\omega,\alpha,\beta}) = \lim_{\omega \downarrow 0} \kappa(\Omega_{\omega,\alpha,\beta}) = \left(\sqrt{1+\alpha^2} - \sqrt{1-\beta^2}\right) / \sqrt{\alpha^2+\beta^2}$.

(ii) *For $\omega_2 \le \omega \le \omega_3$, $\kappa(\Omega_{\omega,\alpha,\beta})$ is a constant function of ω which satisfies*

$$\min\{1, \rho(\mathcal{L}_\omega)\} > \kappa(\Omega_{\omega,\alpha,\beta}) = \frac{\left(\sqrt{1+\alpha^2} - \sqrt{1-\beta^2}\right)^2}{\alpha^2 + \beta^2}. \tag{17}$$

(iii) *For $\omega_3 < \omega < \omega_4$, $\kappa(\Omega_{\omega,\alpha,\beta})$ is a strictly increasing function of ω which satisfies*

$$\min\{1, \rho(\mathcal{L}_\omega)\} > \kappa(\Omega_{\omega,\alpha,\beta}) > \frac{\left(\sqrt{1+\alpha^2} - \sqrt{1-\beta^2}\right)^2}{\alpha^2 + \beta^2}. \tag{18}$$

(iv) *In all remaining cases, i.e., for $-\infty < \omega \le \omega_1$, for $\omega_4 \le \omega < \infty$, and for $\omega = 0$,*

$$\min\{1, \rho(\mathcal{L}_\omega)\} = \kappa(\Omega_{\omega,\alpha,\beta}) = 1, \tag{19}$$

i.e., neither the SOR method (5) nor any semi-iterative method applied to this SOR method can converge for these values of ω.

As a consequence of the above,

$$1 > \min_{\omega \in \mathbb{R}} \rho(\mathcal{L}_\omega) > \min_{\omega \in \mathbb{R}} \kappa(\Omega_{\omega,\alpha,\beta}) = \frac{\left(\sqrt{1+\alpha^2} - \sqrt{1-\beta^2}\right)^2}{\alpha^2 + \beta^2}. \tag{20}$$

The results of Theorem 2 can be seen in Figure 2 for $\alpha = 1$ and $\beta = 0.9$. For these choices of α and β, the relevant quantities of (8) and (9) are

$$\omega_b = 0.95653\ldots \text{ and } \rho(\mathcal{L}_{\omega_b}) = 0.82575\ldots,$$

the relevant quantities of (14) are

$$\omega_1 = -4.82842\ldots, \quad \omega_2 = 0.82842\ldots, \quad \omega_3 = 1.39286\ldots, \text{ and } \omega_4 = 3.54540\ldots,$$

while from (17), we have

$$\kappa(\Omega_{\omega,\alpha,\beta}) = 0.52879\ldots \text{ for all } \omega \in [\omega_2, \omega_3].$$

We again emphasize that the main result of Theorem 2 (cf. (20)),

$$\min_{\omega \in \mathbf{R}} \rho(\mathcal{L}_\omega) > \min_{\omega \in \mathbf{R}} \kappa(\Omega_{\omega,\alpha,\beta}) \tag{21}$$

crucially depends on the above assumptions that $\alpha > 0$ *and* $\beta > 0$. In [5, Theorem 1], we have shown that equality holds in (21) for the nonnegative case, i.e., for $\alpha = 0$ or, equivalently, $\sigma(B^2) \subset [0, \beta^2]$. In Theorem 1, we saw that equality also holds in (21) for the nonpositive case, i.e., for $\beta = 0$ or, equivalently, $\sigma(B^2) \subset [-\alpha^2, 0]$.

We add another remark. Instead of applying a semi-iterative method to the SOR iteration as we did in (10), we could have applied a semi-iterative method directly to the Jacobi method

$$\mathbf{x}_m = B\mathbf{x}_{m-1} + D^{-1}\mathbf{b} \qquad (m = 1, 2, \ldots). \tag{22}$$

From $\sigma(B) \subset [-\alpha^2, \beta^2]$, we conclude that

$$\sigma(B) \subset [-\beta, \beta] \cup [-i\alpha, i\alpha],$$

and one might ask what is the smallest asymptotic convergence factor $\kappa([-\beta, \beta] \cup [-i\alpha, i\alpha])$ of any semi-iterative method applied to (22) with respect to this information. It turns out (cf. [2, eq. (4.6)]) that

$$\kappa([-\beta, \beta] \cup [-i\alpha, i\alpha]) = \frac{\sqrt{1+\alpha^2} - \sqrt{1-\beta^2}}{\sqrt{\alpha^2 + \beta^2}}, \tag{23}$$

and consequently (cf. (17)), the optimal semi-iterative method applied to the SOR iterative method with $\omega = 1$ is exactly *twice* as fast as the optimal semi-iterative method applied to the Jacobi iteration!

4 Covering domains $\Omega_{\omega,\alpha,\beta}$ and their asymptotic convergence factors

From our initial assumption that the Jacobi matrix B of (2) is consistently ordered and weakly cyclic of index 2, then Young's fundamental relationship

$$(\lambda + \omega - 1)^2 = \lambda \omega^2 \mu^2 \tag{24}$$

holds between the eigenvalues λ of $\mathcal{L}_\omega$ and the eigenvalues of μ of B (cf. [10] or [11, Theorem 5-2.2]). The relation (24) will be used to determine sharp covering domains $\Omega_{\omega,\alpha,\beta}$ for the eigenvalues of $\mathcal{L}_\omega$. In analogy with [11, pp. 203-206], it is necessary to distinguish between several cases. To this end, we first define certain "extremal" eigenvalues of $\mathcal{L}_\omega$, namely

$$\begin{aligned} \lambda_1 &= \lambda_1(\omega, \beta) := \left[\frac{\omega\beta - \sqrt{\omega^2\beta^2 - 4(\omega-1)}}{2}\right]^2, \\ \lambda_2 &= \lambda_2(\omega, \beta) := \left[\frac{\omega\beta + \sqrt{\omega^2\beta^2 - 4(\omega-1)}}{2}\right]^2 \end{aligned} \tag{25}$$

(for given real ω and $\mu^2 = \beta^2$, λ_1 and λ_2 are just the two roots of (24)), and

$$\begin{aligned} \lambda_3 &= \lambda_3(\omega,\alpha) := -\left[\frac{\omega\alpha - \sqrt{\omega^2\alpha^2 + 4(\omega-1)}}{2}\right]^2, \\ \lambda_4 &= \lambda_4(\omega,\alpha) := -\left[\frac{\omega\alpha + \sqrt{\omega^2\alpha^2 + 4(\omega-1)}}{2}\right]^2 \end{aligned} \tag{26}$$

(for given real ω and $\mu^2 = -\alpha^2$, λ_3 and λ_4 are again just the two roots of (24)). For these extremal eigenvalues of $\mathcal{L}_\omega$, it can be verified that

$$\begin{aligned} &\lambda_1 \text{ and } \lambda_2 \text{ are both nonreal if and only if } \omega \in (\omega_3, \omega_4)\,, \text{ and} \\ &\lambda_3 \text{ and } \lambda_4 \text{ are both nonreal if and only if } \omega \in (\omega_1, \omega_2) \setminus \{0\}\,. \end{aligned}$$

For any real ω, it follows from (14) that ω necessarily lies in one of the five disjoint real intervals: $-\infty < \omega \le \omega_1$, $\omega_1 < \omega < \omega_2$, $\omega_2 \le \omega \le \omega_3$, $\omega_3 < \omega < \omega_4$, and $\omega_4 \le \omega < \infty$.

Our immediate goal below is to determine sharp *covering domains* $\Omega_{\omega,\alpha,\beta}$ for the spectrum, $\sigma(\mathcal{L}_\omega)$, of the SOR iterative matrix $\mathcal{L}_\omega$, as a function of ω in these five intervals.

Case 1: $\omega \in (-\infty, \omega_1]$. Fixing any ω in $(-\infty, \omega_1]$, the image of the two roots λ of (24), for this fixed ω and for a variable μ^2 satisfying $0 \le \mu^2 \le \beta^2$, can be verified to be the real interval $[\lambda_2, \lambda_1]$, where λ_1 and λ_2 are given in (25). In the same fashion, the image of the two roots λ of (24), for a fixed ω and a variable μ^2 satisfying $-\alpha^2 \le \mu^2 \le 0$, can be verified, with (26), to be the union of the real interval $[\lambda_4, \lambda_3]$ and the circle $\partial\mathbb{D}(0; 1-\omega)$, where $\partial\mathbb{D}(a;b) := \{z \in \mathbb{C} : |z-a| = b\}$. More geometrically, for $\mu^2 = \beta^2$ these images are just λ_1 and λ_2 from (25), and decreasing μ^2 from β^2 moves these λ's toward one another until these images meet in the common point $1-\omega$ when $\mu^2 = 0$. On decreasing μ^2 further for $-\alpha^2 \le \mu^2 \le 0$, these images move, as conjugate complex pairs, along the circle $\partial\mathbb{D}(0; 1-\omega)$ until they meet in the point $\omega - 1$, where they separate and trace out the real interval $[\lambda_3, \lambda_4]$, where λ_3 and λ_4 are defined in (26). (This movement of the λ's, as μ^2 decreases from β^2 to $-\alpha^2$, is indicated by the arrows in Figure 3a.) Thus, from Young's fundamental relationship (24), the spectrum $\sigma(\mathcal{L}_\omega)$, for any ω in $(-\infty, \omega_1]$, satisfies

$$\begin{aligned} &\sigma(\mathcal{L}_\omega) \subset \Omega_{\omega,\alpha,\beta} := [\lambda_3, \lambda_4] \cup \partial\mathbb{D}(0; 1-\omega) \cup [\lambda_2, \lambda_1] \\ &\text{with } \lambda_3 \le \omega - 1 \le \lambda_4 < 1 < \lambda_2 < 1 - \omega < \lambda_1\,, \end{aligned}$$

and this $\Omega_{\omega,\alpha,\beta}$ (cf. Figure 3a) is the associated *covering domain* for $\sigma(\mathcal{L}_\omega)$.

We note in this case that $1 \notin \Omega_{\omega,\alpha,\beta}$ since $\omega - 1 \le \lambda_4 < 1 < \lambda_2 < 1 - \omega$. However (cf. (12)), $\Omega_{\omega,\alpha,\beta} \notin \mathbb{M}$ because the complement of $\Omega_{\omega,\alpha,\beta}$ is clearly not connected. Moreover, since 1 and ∞ are contained in different components of $C_\infty \setminus \Omega_{\omega,\alpha,\beta}$, we conclude from the discussion following (11) that

$$\kappa(\Omega_{\omega,\alpha,\beta}) = 1 \qquad (-\infty < \omega \le \omega_1)\,. \tag{27}$$

Consequently, *no* semi-iterative method applied to the SOR iterative method converges for *any* ω in this interval $(-\infty, \omega_1]$.

We further remark that the above covering domain $\Omega_{\omega,\alpha,\beta}$ is also the limiting case of the covering domain of Case 2 when $\omega \downarrow \omega_1$. (This continuity of the covering domains, in passing from one interval in ω to the next, is valid in all cases considered below.)

Case 2: $\omega \in (\omega_1, \omega_2)$ **and** $\omega \neq \mathbf{0}$. Fixing any $\omega \neq 0$ in (ω_1, ω_2), the use of (24), as in Case 1, similarly determines a covering domain $\Omega_{\omega,\alpha,\beta}$ for $\sigma(\mathcal{L}_\omega)$. Specifically, it can be verified that if $\omega \in (\omega_1, \omega_2)$ with $\omega < 0$, then

$$\begin{aligned} &\sigma(\mathcal{L}_\omega) \subset \Omega_{\omega,\alpha,\beta} := \{z : |z| = 1-\omega \text{ and } |\arg z| \leq \arg \lambda_4\} \cup [\lambda_2, \lambda_1] \\ &\text{with } 1 < \lambda_2 < 1-\omega < \lambda_1 , \end{aligned} \tag{28}$$

while if $\omega \in (\omega_1, \omega_2)$ with $\omega > 0$, then

$$\begin{aligned} &\sigma(\mathcal{L}_\omega) \subset \Omega_{\omega,\alpha,\beta} := \{z : |z| = 1-\omega \text{ and } |\arg z| \leq \arg \lambda_3\} \cup [\lambda_1, \lambda_2] \\ &\text{with } \lambda_1 < 1-\omega < \lambda_2 < 1 . \end{aligned} \tag{29}$$

For either $\omega < 0$ or $\omega > 0$ of this case, $\Omega_{\omega,\alpha,\beta}$ has the form

$$D_{\tau,\theta,\zeta,\eta} := \left\{\tau e^{i\varphi} : -\theta \leq \varphi \leq \theta\right\} \cup [\zeta, \eta], \tag{30}$$

with $\tau > 0$ and $0 < \theta < \pi$, and with either $\zeta < \tau < \eta < 1$ or $1 < \zeta < \tau < \eta$. In either situation, we see that $1 \notin D_{\tau,\theta,\zeta,\eta}$ and that (cf. Figure 3b) $\mathbb{C}_\infty \setminus D_{\tau,\theta,\zeta,\eta}$ is simply connected. Consequently,

$$D_{\tau,\theta,\zeta,\eta} \in \mathbb{M}.$$

(Its asymptotic convergence factor $\kappa(D_{\tau,\theta,\zeta,\eta})$ will be obtained later in closed form in Proposition 5.)

Case 3: $\omega \in [\omega_2, \omega_3]$. For any $\omega \in [\omega_2, \omega_3]$, it can be similarly verified that

$$\begin{aligned} &\sigma(\mathcal{L}_\omega) \subset \Omega_{\omega,\alpha,\beta} := [\lambda_4, \lambda_3] \cup \partial\mathbb{D}(0; |1-\omega|) \cup [\lambda_1, \lambda_2] \\ &\text{with } \lambda_4 \leq -|\omega - 1| \leq \lambda_3 \leq 0 \leq \lambda_1 \leq |\omega - 1| \leq \lambda_2 < 1 , \end{aligned} \tag{31}$$

where the associated covering domain $\Omega_{\omega,\alpha,\beta}$ is shown in Figure 3c.

Since $|1-\omega| \leq \lambda_2 < 1$ from (31), the critical point $z = 1$ is now not in $\Omega_{\omega,\alpha,\beta}$. In addition (cf. Figure 3c), the intervals $(-|1-\omega|, \lambda_3]$ and $[\lambda_1, |1-\omega|)$, which are *interior* to the closed disk $\overline{\mathbb{D}}(0; |1-\omega|)$, have no effect on the determination of the asymptotic convergence factor of $\Omega_{\omega,\alpha,\beta}$, as can be seen from applying the maximum principle to the expression in (11). Hence, the new covering domain in this case, with the identical asymptotic convergence factor, is of the form

$$B_{\tau,\zeta,\eta} := \overline{\mathbb{D}}(0; \tau) \cup [\zeta, \eta] , \quad \text{where } \zeta \leq -\tau < \tau \leq \eta < 1, \tag{32}$$

with the choices $\tau = |1-\omega|, \zeta = \lambda_4$ and $\eta = \lambda_2$ (cf. Figure 3d). Now, $B_{\tau,\zeta,\eta} \in \mathbb{M}$ so that

$$\kappa(\Omega_{\omega,\alpha,\beta}) = \kappa(B_{\tau,\zeta,\eta}) , \quad \text{where } \tau := |1-\omega|, \zeta := \lambda_4 \text{ and } \eta := \lambda_2. \tag{33}$$

(The asymptotic convergence factor $\kappa(B_{\tau,\zeta,\eta})$ will be obtained later in closed form in Proposition 3.)

We remark that if $\omega = 1$ (i.e., SOR is the Gauss–Seidel method), then from (25) and (26),

$$\lambda_1 = \lambda_3 = 0\,, \ \ \lambda_4 = -\alpha^2\,, \text{ and } \lambda_2 = \beta^2,$$

so that $\Omega_{1,\alpha,\beta}$ reduces in this case to the interval $[-\alpha^2, \beta^2]$. For the extremal values of ω for Case 3 (i.e., ω_2 and ω_3), we also remark that a similar reduction takes place, in that it can be verified that

$$\omega = \omega_2 \text{ implies } \lambda_3 = \lambda_4 = \omega_2 - 1 \text{ and } \Omega_{\omega_2,\alpha,\beta} = \partial\mathbf{D}(0; 1-\omega_2) \cup [\lambda_1, \lambda_2]\,,$$

and

$$\omega = \omega_3 \text{ implies } \lambda_1 = \lambda_2 = \omega_3 - 1 \text{ and } \Omega_{\omega_3,\alpha,\beta} = [\lambda_4, \lambda_3] \cup \partial\mathbf{D}(0; \omega_3 - 1)\,.$$

Case 4: $\omega_3 < \omega < \omega_4$. Fixing any ω in (ω_3, ω_4), it can be verified that

$$\sigma(\mathcal{L}_\omega) \subset \Omega_{\omega,\alpha,\beta} := \{z : |z| = \omega - 1 \text{ and } \arg\lambda_2 \le \arg z \le \arg\lambda_1\} \cup [\lambda_4, \lambda_3]$$
$$\text{with } \operatorname{Im}\lambda_2 > 0 \text{ and } \lambda_4 < 1 - \omega < \lambda_3 < 0\,.$$

With the choice $\tau = |\omega - 1|$, $\theta = \arg\lambda_2$, $\zeta = \lambda_4$, and $\eta = \lambda_3$, the associated covering domain has the form

$$C_{\tau,\theta,\zeta,\eta} := \left\{\tau e^{i\varphi} : \theta \le \varphi \le 2\pi - \theta\right\} \cup [\zeta, \eta], \tag{34}$$

where $\tau > 0, 0 < \theta < \pi$ and $\zeta < -\tau < \eta < 1$. This is shown in Figure 3e. Again, $1 \notin C_{\tau,\theta,\zeta,\eta}$, and $C_\infty \setminus C_{\tau,\theta,\zeta,\eta}$ is simply connected, so that

$$C_{\tau,\theta,\zeta,\eta} \in \mathbf{M}.$$

(Its convergence factor $\kappa(C_{\tau,\theta,\zeta,\eta})$ will be obtained later in closed form in Proposition 4.)

Case 5: $\omega_4 \le \omega < \infty$. Fixing any ω in $[\omega_4, \infty)$, it can be verified that

$$\sigma(\mathcal{L}_\omega) \subset \Omega_{\omega,\alpha,\beta} := [\lambda_4, \lambda_3] \cup \partial\mathbf{D}(0; \omega - 1) \cup [\lambda_1, \lambda_2]$$
$$\text{with } \lambda_4 < 1 - \omega < \lambda_3 < 1 < \lambda_1 \le \omega - 1 \le \lambda_2\,,$$

which is shown in Figure 3f. Now, $1 \notin \Omega_{\omega,\alpha,\beta}$, but as in Case 1, since the complement of $\Omega_{\omega,\alpha,\beta}$ is not connected with 1 and ∞ in different components, we again have, from the discussion following (11), that

$$\kappa(\Omega_{\omega,\alpha,\beta}) = 1 \qquad (\omega_4 \le \omega < \infty)\,, \tag{35}$$

i.e., no semi-iterative method applied to the SOR iterative method converges for any ω in $[\omega_4, \infty)$.

With the above five cases for the determination of the associated covering domains $\Omega_{\omega,\alpha,\beta}$, only Cases 2-4 have any interest for us since the remaining two (cf. (27) and (35)) result in no convergence via semi-iterative methods for the SOR iterative method. But in Cases 2-4 (i.e., $D_{\tau,\theta,\zeta,\eta}$ for Case 2, $B_{\tau,\zeta,\eta}$ for Case 3, and

$C_{\tau,\theta,\zeta,\eta}$ for Case 4), each has an associated covering domain in $\mathbb{M}$ from which, using (13), its asymptotic convergence factor can be determined. We now explicitly determine the asymptotic convergence factors for the three cases. We begin with the simplest case of

Proposition 3. *The asymptotic convergence factor $\kappa(B_{\tau,\zeta,\eta})$ of the set $B_{\tau,\zeta,\eta}$ of (32) is given by*

$$\kappa\left(B_{\tau,\zeta,\eta}\right) = t - \sqrt{t^2 - 1}\,,$$

where

$$t := \frac{(\zeta+\eta)(\tau^2+\zeta\eta) - 2\zeta\eta(\tau^2+1)}{(\eta-\zeta)(\tau^2-\zeta\eta)}\ (>1)\,, \tag{36}$$

provided that $\zeta \leq -\tau < \tau \leq \eta < 1$.

Proof. Since, as noted above, $B_{\tau,\zeta,\eta} \in \mathbb{M}$, then (13) can be applied to determine $\kappa\left(B_{\tau,\zeta,\eta}\right)$. To this end, we explicitly construct a conformal mapping function Φ which maps $\mathbb{C}_\infty \setminus B_{\tau,\zeta,\eta}$ onto $\mathbb{C}_\infty \setminus \overline{\mathbb{D}}(0;1)$, where Φ is normalized by $\Phi(\infty) = \infty$. The mapping Φ can be expressed as the composition of three elementary mappings. The Joukowski-like transformation

$$u = \Phi_1(z) := \frac{z}{\tau} + \frac{\tau}{z}$$

maps $\mathbb{C}_\infty \backslash B_{\tau,\zeta,\eta}$ in the z-plane conformally onto $\mathbb{C}_\infty \backslash [\Phi_1(\zeta), \Phi_1(\eta)]$ in the u-plane, the linear transformation

$$v = \Phi_2(u) := \frac{2u - \Phi_1(\zeta) - \Phi_1(\eta)}{\Phi_1(\eta) - \Phi_1(\zeta)}$$

maps $\mathbb{C}_\infty \setminus [\Phi_1(\zeta), \Phi_1(\eta)]$ in the u-plane conformally onto $\mathbb{C}_\infty \setminus [-1,1]$ in the v-plane, and, finally, the inverse Joukowski transformation

$$w = \Phi_3(v) := v + \sqrt{v^2 - 1},$$

maps $\mathbb{C}_\infty \setminus [-1,1]$ in the v-plane conformally onto $\mathbb{C}_\infty \setminus \overline{\mathbb{D}}(0;1)$ in the w-plane, where the branch of the square root is chosen so that $\Phi_3(v) > 1$ for all $v > 1$. (This choice guarantees that $|\Phi_3(v)| > 1$ for all $v \notin [-1,+1]$.)

The composition of these mappings, namely $\Phi = \Phi_3 \circ \Phi_2 \circ \Phi_1$, then maps $\mathbb{C}_\infty \setminus B_{\tau,\zeta,\eta}$ conformally onto $\mathbb{C}_\infty \setminus \overline{\mathbb{D}}(0;1)$ with $\Phi(\infty) = \infty$, and as $B_{\tau,\zeta,\eta} \in \mathbb{M}$, it follows from (13) that

$$\kappa\left(B_{\tau,\zeta,\eta}\right) = \frac{1}{|\Phi(1)|} = \frac{1}{|(\Phi_3 \circ \Phi_2 \circ \Phi_1)(1)|}$$

has the desired form given in Proposition 3. □

Next, we determine the asymptotic convergence factor $\kappa\left(C_{\tau,\theta,\zeta,\eta}\right)$ for the set $C_{\tau,\theta,\zeta,\eta}$ of (34), which is also an element of $\mathbb{M}$. This determination again makes use of (13), as well as Proposition 3.

Proposition 4. *The asymptotic convergence factor* $\kappa(C_{\tau,\theta,\zeta,\eta})$ *of the set* $C_{\tau,\theta,\zeta,\eta}$ *of* (34) *with* $\tau > 0, 0 < \theta < \pi$ *and* $\zeta < -\tau < \eta < 1$, *satisfies*

$$\kappa(C_{\tau,\theta,\zeta,\eta}) = \kappa(B_{\hat{\tau},\hat{\zeta},\hat{\eta}}), \tag{37}$$

where $B_{\hat{\tau},\hat{\zeta},\hat{\eta}}$ *is defined from* (32) *with*

$$\hat{\tau} := 2\cos\left(\frac{\theta}{2}\right)\frac{\tau}{1+\tau+\sqrt{1-2\tau\cos(\theta)+\tau^2}},$$

$$\hat{\zeta} := \frac{\zeta+\tau-\sqrt{\zeta^2-2\zeta\tau\cos(\theta)+\tau^2}}{1+\tau+\sqrt{1-2\tau\cos(\theta)+\tau^2}} \quad \textit{and} \quad \hat{\eta} := \frac{\eta+\tau+\sqrt{\eta^2-2\eta\tau\cos(\theta)+\tau^2}}{1+\tau+\sqrt{1-2\tau\cos(\theta)+\tau^2}}.$$

Proof. We explicitly construct a conformal map Υ of $\mathbb{C}_\infty \backslash C_{\tau,\theta,\zeta,\eta}$ onto $\mathbb{C}_\infty \backslash B_{\hat{\tau},\hat{\zeta},\hat{\eta}}$ with $\Upsilon(\infty) = \infty$ and $\Upsilon(1) = 1$. In view of (13), this will give the assertion of Proposition 4.

As similarly in [5, Proposition 4], we consider the three elementary conformal mappings

$$\begin{aligned} u = \Upsilon_1(z) &:= i\left[\frac{z-\tau\cos(\theta)}{\tau\sin(\theta)}\right], \\ v = \Upsilon_2(u) &:= u+\sqrt{u^2-1}, \text{ and} \\ w = \Upsilon_3(v) &:= \cos\left(\frac{\theta}{2}\right) - i\sin\left(\frac{\theta}{2}\right)v. \end{aligned}$$

A straight-forward computation shows that the composition of these three maps can be expressed as

$$\tilde{\Upsilon}(z) := (\Upsilon_3 \circ \Upsilon_2 \circ \Upsilon_1)(z) = \frac{z+\tau+\sqrt{z^2-2\tau z\cos(\theta)+\tau^2}}{2\tau\cos(\theta/2)}. \tag{38}$$

In [5, Proposition 4], we showed that $\tilde{\Upsilon}$ maps $\mathbb{C}_\infty \setminus \{\tau e^{i\varphi} : \theta \le \varphi \le 2\pi - \theta\}$ conformally onto $\mathbb{C}_\infty \setminus \overline{\mathbb{D}}(0;1)$ with $\tilde{\Upsilon}(\infty) = \infty$, provided that we choose the branch of the square root in (38) for which $\tilde{\Upsilon}(1) > 0$. Now, with $\hat{\tau} := 1/\tilde{\Upsilon}(1) > 0$ (which, from(38), agrees with the definition of $\hat{\tau}$ in Proposition 4),

$$\Upsilon(z) := \hat{\tau}\tilde{\Upsilon}(z)$$

is then the conformal map from $\mathbb{C}_\infty \setminus \{\tau e^{i\varphi} : \theta \le \varphi \le 2\pi - \theta\}$ onto $\mathbb{C}_\infty \setminus \overline{\mathbb{D}}(0;\hat{\tau})$ with $\Upsilon(\infty) = \infty$ and $\Upsilon(1) = 1$. Moreover, $[\zeta, -\tau) \cup (\tau, \eta]$ is mapped by Υ onto $[\hat{\zeta}, -\hat{\tau}) \cup (\hat{\tau}, \hat{\eta}]$ (whose definitions are given in Proposition 4), and it can be verified, from the assumptions on τ, θ, ζ, and η in Proposition 4, that $\hat{\zeta} < -\hat{\tau} < \hat{\tau} < \hat{\eta} < 1$. Consequently, Υ is a conformal map of $\mathbb{C}_\infty \backslash C_{\tau,\theta,\zeta,\eta}$ onto $\mathbb{C}_\infty \backslash B_{\hat{\tau},\hat{\zeta},\hat{\eta}}$ with $\Upsilon(1) = 1$. Appealing again to (13), we conclude that $\kappa(C_{\tau,\theta,\zeta,\eta}) = \kappa(B_{\hat{\tau},\hat{\zeta},\hat{\eta}})$, which is the desired result (37) of Proposition 4. □

Next, the sets $C_{\tau,\theta,\zeta,\eta}$ and $D_{\tau,\theta,\zeta,\eta}$, from Figures 3b and 3e, appear to be simply "reflections" of one another. This simplifies the proof of a

Proposition 5. *The asymptotic convergence factor $\kappa(D_{\tau,\theta,\zeta,\eta})$ of the set $D_{\tau,\theta,\zeta,\eta}$ of (30), with $\tau > 0, 0 < \theta < \pi$ and $\zeta < \tau < \eta < 1$ or $1 < \zeta < \tau < \eta$, satisfies*

$$\kappa(D_{\tau,\theta,\zeta,\eta}) = \kappa(B_{\check{\tau},\check{\zeta},\check{\eta}}), \tag{39}$$

where $B_{\check{\tau},\check{\zeta},\check{\eta}}$ is defined from (32) with

$$\check{\tau} := 2\sin\left(\frac{\theta}{2}\right)\frac{\tau}{1-\tau+\sqrt{1-2\tau\cos(\theta)+\tau^2}},$$

$$\check{\zeta} := \frac{\zeta-\tau-\sqrt{\zeta^2-2\zeta\cos(\theta)+\tau^2}}{1-\tau+\sqrt{1-2\tau\cos(\theta)+\tau^2}} \quad \text{and} \quad \check{\eta} := \frac{\eta-\tau+\sqrt{\eta^2-2\eta\tau\cos(\theta)+\tau^2}}{1-\tau+\sqrt{1-2\tau\cos(\theta)+\tau^2}},$$

provided that $\zeta < \tau < \eta < 1$, and with

$$\check{\tau} := 2\sin\left(\frac{\theta}{2}\right)\frac{\tau}{-1+\tau+\sqrt{1-2\tau\cos(\theta)+\tau^2}},$$

$$\check{\zeta} := \frac{-\eta+\tau-\sqrt{\eta^2-2\eta\tau\cos(\theta)+\tau^2}}{-1+\tau+\sqrt{1-2\tau\cos(\theta)+\tau^2}} \quad \text{and} \quad \check{\eta} := \frac{-\zeta+\tau+\sqrt{\zeta^2-2\zeta\tau\cos(\theta)+\tau^2}}{-1+\tau+\sqrt{1-2\tau\cos(\theta)+\tau^2}},$$

provided that $1 < \zeta < \tau < \eta$.

Proof. This is a direct consequence of Proposition 4 because $z \mapsto -z$ maps $\mathbb{C}_\infty \setminus D_{\tau,\theta,\zeta,\eta}$ onto $\mathbb{C}_\infty \setminus C_{\tilde{\tau},\tilde{\theta},\tilde{\zeta},\tilde{\eta}}$, where $\tilde{\tau} := \tau$, $\tilde{\theta} := \pi - \theta$, $\tilde{\zeta} := -\eta$, and $\tilde{\eta} := -\zeta$. □

As mentioned at the end of Section 2, there are essentially only *three* different types of covering domain $\Omega_{\omega,\alpha,\beta}$ which needed to be analyzed. The asymptotic convergence factors of these three different covering domains now have been determined in the three Propositions 3-5.

Also mentioned in Section 2 is that *sharp* covering domains will be derived in this section. To demonstrate this point, fix any α and β with $0 < \alpha < \infty$ and $0 < \beta < 1$, and fix any real ω with $\omega \neq 0$. Next, let λ be any real or complex number such that if (cf. (24))

$$(\lambda+\omega-1)^2 = \lambda\omega^2\mu^2, \quad \text{then } \mu^2 \text{ satisfies } -\alpha^2 \le \mu^2 \le \beta^2. \tag{40}$$

By definition, λ is then an arbitrary point of the covering domain $\Omega_{\omega,\alpha,\beta}$. Next, with a value of μ (possibly complex), determined by (40) from the given values of λ and ω, we define the matrix $\tilde{B}$ by

$$\tilde{B} := \left[\begin{array}{c|c|c} \begin{matrix} 0 & \beta \\ \beta & 0 \end{matrix} & O & O \\ \hline O & \begin{matrix} 0 & \mu \\ \mu & 0 \end{matrix} & O \\ \hline O & O & \begin{matrix} 0 & \alpha \\ -\alpha & 0 \end{matrix} \end{array}\right], \tag{41}$$

whose eigenvalues are $\pm\beta$, $\pm\mu$, and $\pm i\alpha$. It can be verified, from the case $p = 2$ of [9, Definition 4.2 and Theorem 4.1] that $\tilde{B}$ is a consistently ordered weakly cyclic of index 2 matrix. Moreover, since the diagonal entries of $\tilde{B}$ of (41) are all zero, then on setting $\tilde{B} := \tilde{L} + \tilde{U}$, where $\tilde{L}$ and $\tilde{U}$ are respectively the strictly lower and strictly upper triangular matrix determined from $\tilde{B}$, an associated 6×6 SOR matrix $\tilde{\mathcal{L}}_\omega$ for $\tilde{B}$ can be defined by

$$\tilde{\mathcal{L}}_\omega := (I - \omega\tilde{L})^{-1}[(1-\omega)I + \omega\tilde{U}].$$

But as μ is an eigenvalue of $\tilde{B}$ and as (24) is satisfied, it follows (cf. [9, Theorem 4.3]) that λ is an eigenvalue of $\tilde{\mathcal{L}}_\omega$. In other words, for any $\omega \neq 0$ and for any α and β with $0 < \alpha < \infty$ and $0 < \beta < 1$, each point of the covering domain $\Omega_{\omega,\alpha,\beta}$ *is* an eigenvalue of some SOR matrix derived from a consistently ordered weakly cyclic of index 2 Jacobi matrix B whose spectrum satisfies $\sigma(B^2) \subset [-\alpha^2, \beta^2]$. It is in this sense that the covering domains $\Omega_{\omega,\alpha,\beta}$, for $\omega \neq 0$, are *sharp*.

5 Proof of Theorem 2

With the results of Propositions 3–5, we are in position to establish Theorem 2.
Proof. As seen from the discussion in Section 3, there were five real intervals in ω to be separately considered, namely Cases 1–5. As we have seen (cf. (27) and (35)), Cases 1 and 5 both give an asymptotic convergence factor $\kappa(\Omega_{\omega,\alpha,\beta}) = 1$, so that no semi-iterative method applied to the SOR iterative method is convergent when $-\infty < \omega \leq \omega_1$ (Case 1) and $\omega_4 \leq \omega < \infty$ (Case 5).

For Case 2 (i.e., $\omega \in (\omega_1, \omega_2)$ and $\omega \neq 0$), we distinguish between the two subcases $\omega_1 < \omega < 0$ and $0 < \omega < \omega_2$. Since the treatment of each of these subcases is similar, we consider only the subcase $\omega_1 < \omega < 0$. Then (cf. (28) and the following inequalities),

$$1 < \lambda_2 < \lambda_1\,, \ \mathrm{Im}(\lambda_4) > 0,$$

and $\sigma(\mathcal{L}_\omega) \subset D_{\tau,\theta,\zeta,\eta}$ with $\tau := 1-\omega$, $\theta := \arg(\lambda_4)$, $\zeta := \lambda_2$ and $\eta := \lambda_1$. Inserting these definitions into the expression for $\check{\tau}$, $\check{\zeta}$, and $\check{\eta}$ of Proposition 5, we obtain (after some algebraic manipulations and simplifications) that

$$\check{\tau} := \sqrt{(1-\omega_2)(1-\omega)},$$

$$\check{\zeta} := -\frac{\beta + \sqrt{\alpha^2+\beta^2}}{1+\sqrt{1+\alpha^2}}\sqrt{\lambda_1} \ \text{ and } \ \check{\eta} := \frac{\beta + \sqrt{\alpha^2+\beta^2}}{1+\sqrt{1+\alpha^2}}\sqrt{\lambda_2}\,.$$

Now, these values $\check{\tau}$, $\check{\zeta}$, $\check{\eta}$ determine an associated $\check{t}$ from (36) of Proposition 3, which, after simplifications, can be expressed in terms of ω, α, and β as

$$\check{t} := \frac{(\check{\zeta}+\check{\eta})(\check{\tau}^2+\check{\zeta}\check{\eta}) - 2\check{\zeta}\check{\eta}(\check{\tau}^2+1)}{(\check{\eta}-\check{\zeta})(\check{\tau}^2-\check{\zeta}\check{\eta})} = \frac{(1-\beta^2-\sqrt{1+\alpha^2})\omega + 2\sqrt{1+\alpha^2}}{\sqrt{\alpha^2+\beta^2}\sqrt{\beta^2\omega^2 - 4(\omega-1)}}. \tag{42}$$

Hence, from Propositions 3 and 5, it follows that the asymptotic convergence factor $\kappa(\Omega_{\omega,\alpha,\beta})$ for the covering domain $\Omega_{\omega,\alpha,\beta}$ when $\omega_1 < \omega < 0$ is given by

$$\kappa(\Omega_{\omega,\alpha,\beta}) = \kappa(D_{\tau,\theta,\zeta,\eta}) = \kappa\left(B_{\hat{\tau},\hat{\zeta},\hat{\eta}}\right) = \check{t} - \sqrt{\check{t}^2 - 1}\,, \tag{43}$$

where $\check{t} > 1$ is explicitly given, as a function of ω, α, and β, in the final expression of (42). With α and β fixed with $\alpha > 0$ and $0 < \beta < 1$, differentiation of $\check{t} = \check{t}(\omega)$ with respect to ω yields, after some manipulations, that

$$\frac{d\check{t}}{d\omega} = \frac{2(1-\beta^2)(\sqrt{1+\alpha^2}-1)}{\beta^3\sqrt{\alpha^2+\beta^2}} \cdot \frac{(\omega-\omega_1)}{[(\omega_3-\omega)(\omega_4-\omega)]^{3/2}} > 0 \quad (\omega_1 < \omega < 0)\,, \tag{44}$$

so that for $\omega_1 < \omega < 0$ $(< \omega_3 < \omega_4)$, $\check{t}$ is a strictly increasing function of ω in the interval $(\omega_1, 0)$. On the other hand, as

$$\frac{d}{d\check{t}}\left\{\check{t} - \sqrt{\check{t}^2-1}\right\} = \frac{(\check{t}^2-1)^{1/2} - \check{t}}{\sqrt{\check{t}^2-1}} < 0 \quad (\text{for all } \check{t} > 1)\,, \tag{45}$$

then $\check{t} - \sqrt{\check{t}^2-1}$ is a strictly decreasing function of $\check{t}$ in the interval $[1,\infty)$. Hence (cf. (43)), it follows from (44) and (45) that $\kappa(\Omega_{\omega,\alpha,\beta})$ is a *strictly decreasing* function of ω in the interval $(\omega_1, 0)$, as claimed in (i) of Theorem 2. (This can also be seen in Figure 2.) As mentioned above, the treatment of the remaining subcase of (i) of Theorem 2, namely, $0 < \omega < \omega_2$, is similar and is omitted.

We now turn to case (ii) of Theorem 2, i.e., $\omega \in [\omega_2, \omega_3]$. From our discussion of Case 3 in Section 4, we know from (33) and (13) that

$$\kappa(\Omega_{\omega,\alpha,\beta}) = \kappa(B_{\tau,\zeta,\eta}) \text{ with } \tau := |1-\omega|\,,\ \zeta := \lambda_4\,, \text{ and } \eta := \lambda_2\,, \tag{46}$$

where λ_2 and λ_4, defined in (25) and (26), are both real and nonzero from (31). We specifically treat now the subcase when $1 < \omega \leq \omega_3$, so that $\tau = \omega - 1 > 0$ and λ_2 and λ_4 are both real and nonzero. From (36) of Proposition 3, the associated value for t in (36) (for the values $\tau = \omega - 1$, $\zeta = \lambda_4$ and $\eta = \lambda_2$) is given by

$$t = \frac{2(\frac{1}{\omega-1} + \omega - 1) - (\frac{\zeta}{\omega-1} + \frac{\omega-1}{\zeta} + \frac{\eta}{\omega-1} + \frac{\omega-1}{\eta})}{\frac{\eta}{\omega-1} + \frac{\omega-1}{\eta} - \frac{\zeta}{\omega-1} - \frac{\omega-1}{\zeta}}\,. \tag{47}$$

Since $\zeta = \lambda_4$ and $\eta = \lambda_2$ are particular roots of (24), then

$$\frac{\eta}{\omega-1} + \frac{\omega-1}{\eta} = \frac{\omega^2\beta^2}{\omega-1} - 2 \text{ and } \frac{\zeta}{\omega-1} + \frac{\omega-1}{\zeta} = -\frac{\omega^2\alpha^2}{\omega-1} - 2\,,$$

and substituting these expressions in (47) yields

$$t = \frac{2 - (\beta^2 - \alpha^2)}{\alpha^2 + \beta^2}\,.$$

Thus by Proposition 3,

$$\kappa(\Omega_{\omega,\alpha,\beta}) = t - \sqrt{t^2-1} = \frac{(\sqrt{1+\alpha^2} - \sqrt{1-\beta^2})^2}{\alpha^2+\beta^2}\,, \tag{48}$$

for all $1 < \omega \leq \omega_3$. The case $\omega_2 \leq \omega < 1$ is similar, and gives the same result in (48). For $\omega = 1$, the covering domain $\Omega_{\omega,\alpha,\beta}$ reduces to the interval $[-\alpha^2, \beta^2]$ whose asymptotic convergence factor is known to be

$$\kappa([-\alpha^2, \beta^2]) = \frac{(\sqrt{1+\alpha^2} - \sqrt{1-\beta^2})^2}{\alpha^2+\beta^2}$$

(cf. [2, §6]). This, together with (9), establishes the assertion (17) of Theorem 2.

Next, let ω satisfy $\omega_3 < \omega < \omega_4$ (cf. (14)), which corresponds to Case 4 of Section 4. Note that in this case,

$$\lambda_4 < \lambda_3 < 0 \text{ and } \operatorname{Im}(\lambda_2) > 0$$

(cf. (25) and (26)). Thus, $\sigma(\mathcal{L}_\omega) \subset C_{\tau,\theta,\zeta,\eta}$ (cf. (34)) with $\tau := \omega - 1$, $\theta := \arg(\lambda_2)$, $\zeta := \lambda_4$, and $\eta := \lambda_3$. Inserting into the expressions for $\hat{\tau}$, $\hat{\zeta}$, and $\hat{\eta}$ which are defined in Proposition 4, we obtain similarly

$$\hat{\tau} := \sqrt{(1-\omega_3)(1-\omega)},$$

$$\hat{\zeta} := -\frac{\alpha + \sqrt{\alpha^2+\beta^2}}{1+\sqrt{1-\beta^2}}\sqrt{-\lambda_4} \text{ and } \hat{\eta} := \frac{\alpha + \sqrt{\alpha^2+\beta^2}}{1+\sqrt{1-\beta^2}}\sqrt{-\lambda_3}\,.$$

A further calculation shows that

$$\hat{t} := \frac{(\hat{\zeta}+\hat{\eta})(\hat{\tau}^2+\hat{\zeta}\hat{\eta}) - 2\hat{\zeta}\hat{\eta}(\hat{\tau}^2+1)}{(\hat{\eta}-\hat{\zeta})(\hat{\tau}^2-\hat{\zeta}\hat{\eta})} = \frac{(\alpha^2+1-\sqrt{1-\beta^2})\omega + 2\sqrt{1-\beta^2}}{\sqrt{\alpha^2+\beta^2}\sqrt{\alpha^2\omega^2+4(\omega-1)}}\,. \quad (49)$$

Hence, from Propositions 3 and 4, we have

$$\kappa(\Omega_{\omega,\alpha,\beta}) = \kappa(\mathbb{C}_{\tau,\zeta,\eta,\theta}) = \kappa\left(B_{\hat{\tau},\hat{\zeta},\hat{\eta}}\right) = \hat{t} - \sqrt{\hat{t}^2-1} \quad \text{(where } \hat{t} > 1). \quad (50)$$

Then, using (49), we deduce that

$$\frac{d\hat{t}}{d\omega} = \frac{2(\alpha^2+1)(1-\sqrt{1-\beta^2})}{\sqrt{\alpha^2+\beta^2}}\,\frac{(\omega-\omega_4)}{[(\omega-\omega_1)(\omega-\omega_2)]^{3/2}} < 0 \quad (\omega_3 < \omega < \omega_4), \quad (51)$$

so that $\hat{t}$ is a strictly decreasing function of $\omega \in (\omega_2, \omega_4)$. But as $\hat{t} - \sqrt{\hat{t}^2-1}$ is also a strictly decreasing function of $\hat{t} > 1$ from (45), it follows that $\kappa(\Omega_{\omega,\alpha,\beta})$ of (50) is a then strictly increasing function of ω in (ω_3, ω_4), as claimed in (iii) of Theorem 2. (This can also be seen in Figure 2.)

To complete the proof of Theorem 2, consider the omitted value $\omega = 0$. In this case (cf. (6)), $\mathcal{L}_0 = I_N$, and its associated covering domain is $\Omega_{\omega,\alpha,\beta} = \{1\} \notin \mathbf{M}$. We use (11) to deduce that $\kappa(\Omega_{\omega,\alpha,\beta}) = 1$. Hence, $\kappa(\Omega_{\omega,\alpha,\beta}) = 1 = \rho(\mathcal{L}_0)$, and neither the SOR method for $\omega = 0$, nor any semi-iterative method applied to $\mathcal{L}_0$, converges in this case. □

6 Remarks and conclusions

Theorem 2, the main result of this paper, has a lengthy statement and an even lengthier proof, since this proof depends on treating, via conformal mapping theory, five disjoint real intervals in which the real parameter ω (the relaxation parameter of the SOR iteration method) can traverse. While the proof is somewhat condensed, we hope that the reader has a complete picture of the arguments involved in this proof.

For remarks and conclusions related to Theorem 2, we give the following:

1. In Theorem 2, it is assumed that the eigenvalues of B^2 (where B is the associated Jacobi matrix of (2)) are all real and lie in $[-\alpha^2, \beta^2]$ with $\alpha > 0$ and $0 < \beta < 1$. On letting $\alpha \downarrow 0$, it can be verified that the results of Theorem 1 reduce exactly to the results of Theorem 1 of [5]. As such, Theorem 2 of this paper is obviously a generalization of Theorem 1 of [5]. We remark however again, that, on letting $\alpha \downarrow 0$, equality then holds in (20).

2. Similarly, one can let $\beta \downarrow 0$ (i.e., the eigenvalues of B^2 are all real and lie in $[-\alpha^2, 0]$), so that the result of Theorem 2 can be used to deduce Theorem 1. We further remark that letting $\beta \downarrow 0$ also gives equality in (20).

3. In essence, Theorem 2 of this paper assumes knowledge of the extreme eigenvalues, namely $-\alpha^2$ and β^2, of B^2, whose eigenvalues are assumed to lie in the interval $[-\alpha^2, \beta^2]$ with $\alpha > 0$ and with $0 < \beta < 1$. It is natural to ask here, as in [5, Theorem 5] (see also Dancis [1]), if *improvements* of the asymptotic convergent rates, for optimal semi-iteration applied to the SOR iteration method, are *possible* if one assumes further *explicit* knowledge of the spectrum of B^2, i.e., say, that the k largest eigenvalues and the ℓ smallest eigenvalues of B^2 are assumed known (with $k + \ell > 2$), and the spectrum of B^2 is contained in

$$\bigcup_{j=1}^{\ell-1} \{-\alpha_j^2\} \bigcup_{j=1}^{k-1} \{\beta_j^2\} \cup [-\alpha_\ell^2, \beta_k^2],$$

with $-\alpha_1^2 < -\alpha_2^2 < \cdots < -\alpha_\ell^2 < 0$ and $0 < \beta_k^2 < \beta_{k+1}^2 < \cdots < \beta_1^2 < 1$. The answer is *yes*; one need only apply Theorem 2 of this paper with α^2 and β^2 replaced, respectively, by α_ℓ^2 and β_k^2. (The details of this are easy, and are left to the reader.)

4. We wish to reiterative what is in (20) and what is also clear from Figure 2. Under the assumptions of Theorem 2, the best (for any real ω) asymptotic rate of convergence for semi-iterative methods applied to the SOR iterative method is actually *strictly better* than the best (for any real ω) asymptotic rate of convergence for the associated SOR iterative method.

5. We note that the covering domains $\Omega_{\omega,\alpha,\beta}$ studied in Section 4 were composed only from the "building blocks" of circles (with centers at the origin), their circular arcs (which are symmetric with respect to the real axis), and real intervals. It is worth mentioning that this is a consequence of the fact that Young's relationship (24) is a *quadratic* in λ, with ω and μ^2 real.

6. It is natural to ask how much of this analysis, of optimal semi-iterative methods applied to the SOR matrix, can be extended to the case where B is a consistently ordered weakly cyclic of index p ($p > 2$) matrix. In analogy to (4), we assume that

$$\sigma(B^p) \subset [-\alpha^p, \beta^p] \text{ with } \alpha \geq 0,\ 0 \leq \beta < 1 \text{ and } -\alpha^p, \beta^p \in \sigma(B^p).$$

There is, of course, the known p-cyclic analogue (cf. [9, p. 106]) of (24), namely

$$(\lambda+\omega-1)^p = \lambda^{p-1}\omega^p\mu^p. \tag{52}$$

Since (52) is now a polynomial equation of degree p (> 2) in λ, it is more difficult to find an explicit description of the associated covering domains for $\sigma(\mathcal{L}_\omega)$,

$$\Omega^{(p)}_{\omega,\alpha,\beta} := \left\{\lambda \in \mathbf{C} : (\lambda+\omega-1)^p = \lambda^{p-1}\omega^p\mu^p \text{ for some } \mu^p \in [-\alpha^p, \beta^p]\right\}.$$

For the special case $\omega = 1$, we have

$$\Omega^{(p)}_{1,\alpha,\beta} = [-\alpha^p, \beta^p] \text{ and } \kappa\left(\Omega^{(p)}_{1,\alpha,\beta}\right) = \frac{\left(\sqrt{1+\alpha^p}+\sqrt{1-\beta^p}\right)^2}{\alpha^p+\beta^p}$$

(cf. [2, (4.2)]). This is actually sufficient to show that

$$\min_{\omega\in\mathbf{R}} \rho(\mathcal{L}_\omega) > \min_{\omega\in\mathbf{R}} \kappa\left(\Omega^{(p)}_{\omega,\alpha,\beta}\right) = \kappa\left(\Omega^{(p)}_{1,\alpha,\beta}\right) = \frac{\left(\sqrt{1+\alpha^p}+\sqrt{1-\beta^p}\right)^2}{\alpha^p+\beta^p} \tag{53}$$

holds true in the p-cyclic case for any value of α and β. We briefly sketch the proof of (53): It is well-known that, under the given assumptions, one SOR step is equivalent to p steps of a certain semi-iterative method, the so-called p-step relaxation method, applied to the Jacobi method (see Gutknecht, Niethammer and Varga [6]). Consequently, any semi-iterative method applied to SOR is asymptotically at most p-times faster than an optimal semi-iterative method applied to the Jacobi method. Since the requirement that the eigenvalues of B^p, the pth power of the Jacobi matrix B, are contained in $[-\alpha^p, \beta^p]$ is equivalent to

$$\begin{aligned}\sigma(B) \subset \Sigma_{p,\alpha,\beta} \quad := \quad & \Big\{\mu \in \mathbf{C} : \mu = te^{2k\pi i/p}\ (0 \le t \le \beta) \text{ or} \\ & \mu = se^{(2k+1)\pi i/p}\ (0 \le s \le \alpha) \text{ for } k = 0, 1, \ldots, p-1\Big\},\end{aligned}$$

we conclude that

$$\min_{\omega\in\mathbf{R}} \kappa\left(\Omega^{(p)}_{\omega,\alpha,\beta}\right) \ge \left[\kappa\left(\Sigma_{p,\alpha,\beta}\right)\right]^p.$$

But

$$\left[\kappa\left(\Sigma_{p,\alpha,\beta}\right)\right]^p = \kappa\left([-\alpha^p, \beta^p]\right) = \kappa\left(\Omega^{(p)}_{1,\alpha,\beta}\right)$$

(cf. [2, Theorem 6]) which proves the last two relations of (53). Using the expressions for $\rho(\mathcal{L}_{\omega_b})$ given in [3, §3], an easy calculation finally shows that the first inequality in (53) is also valid.

An interesting open question, which appears to be quite difficult, is whether there exists — as in the two-cyclic case — a whole interval $[\omega_2, \omega_3]$, $\omega_2 < 1 < \omega_3$, of relaxation parameters ω with

$$\kappa\left(\Omega^{(p)}_{\omega,\alpha,\beta}\right) = \min_{\omega\in\mathbf{R}} \kappa\left(\Omega^{(p)}_{\omega,\alpha,\beta}\right) \text{ for all } \omega \in [\omega_2, \omega_3].$$

These problems are of recent interest since they have applications to Markov chains (cf. Kontovasilis, Plemmons, and Stewart [7]).

References

[1] J. Dancis. *The optimal ω is not best for the SOR iteration method.* Linear Algebra Appl. **154-156** (1991), 819-845.

[2] M. Eiermann, X. Li and R. S. Varga. *On hybrid semi-iterative methods.* SIAM J. Numer. Anal. **26** (1989), 152-168.

[3] M. Eiermann, W. Niethammer and A. Ruttan. *Optimal successive overrelaxation methods for p-cyclic matrices.* Numer. Math. **57** (1990), 593-606.

[4] M. Eiermann, W. Niethammer and R. S. Varga. *A study of semi-iterative methods for nonsymmetric systems of linear equations.* Numer. Math. **47** (1985), 505-533.

[5] M. Eiermann and R. S. Varga. *Is the optimal ω best for the SOR iteration method?* Linear Algebra Appl. (to appear).

[6] M. Gutknecht, W. Niethammer and R. S. Varga. *k-step iterative methods for solving nonlinear systems of equations.* Numer. Math. **48** (1986), 699-712.

[7] K. Kontovasilis, R. J. Plemmons and W. J. Stewart *Block cyclic SOR for Markov chains with p-cyclic infinitesimal generator.* Linear Algebra Appl. **154-156** (1991), 145-223.

[8] W. Niethammer. *Relaxation bei Matrizen mit der Eigenschaft "A".* Z. Angew. Math. Mech. **44** (1964), T49-T52.

[9] R. S. Varga. *Matrix Iterative Analysis.* Prentice–Hall, Inc., Englewood Cliffs, N. J., 1962.

[10] D. M. Young. *Iterative methods for solving partial differential equations of elliptic type.* Trans. Amer. Math. Soc. **76** (1954), 92-111.

[11] D. M. Young. *Iterative Solution of Large Linear Systems.* Academic Press, New York 1971.

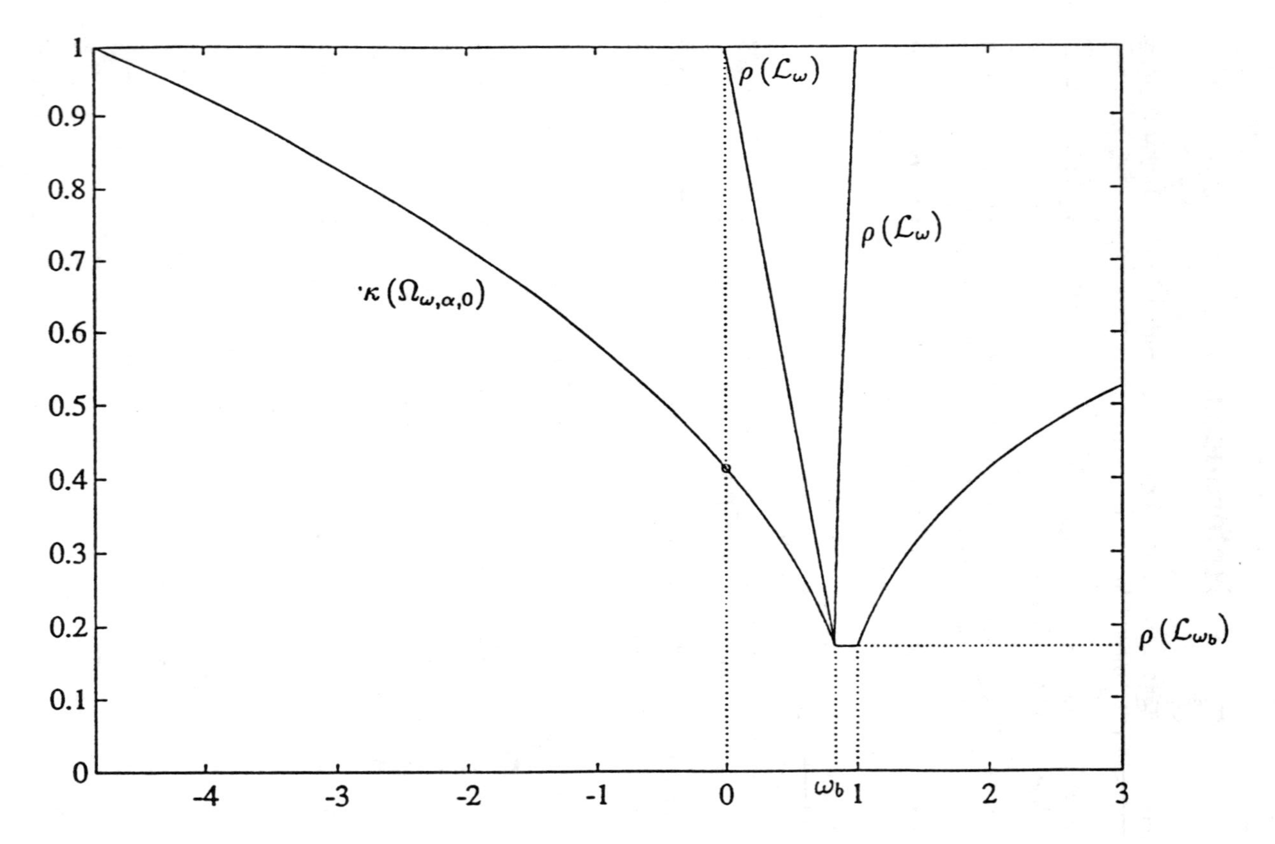

Figure 1: The nonpositive case: $\rho(\mathcal{L}_\omega)$ and $\kappa(\Omega_{\omega,\alpha,0})$ as functions of $\omega \in [\omega_1, 3]$ for $\alpha = 1$.

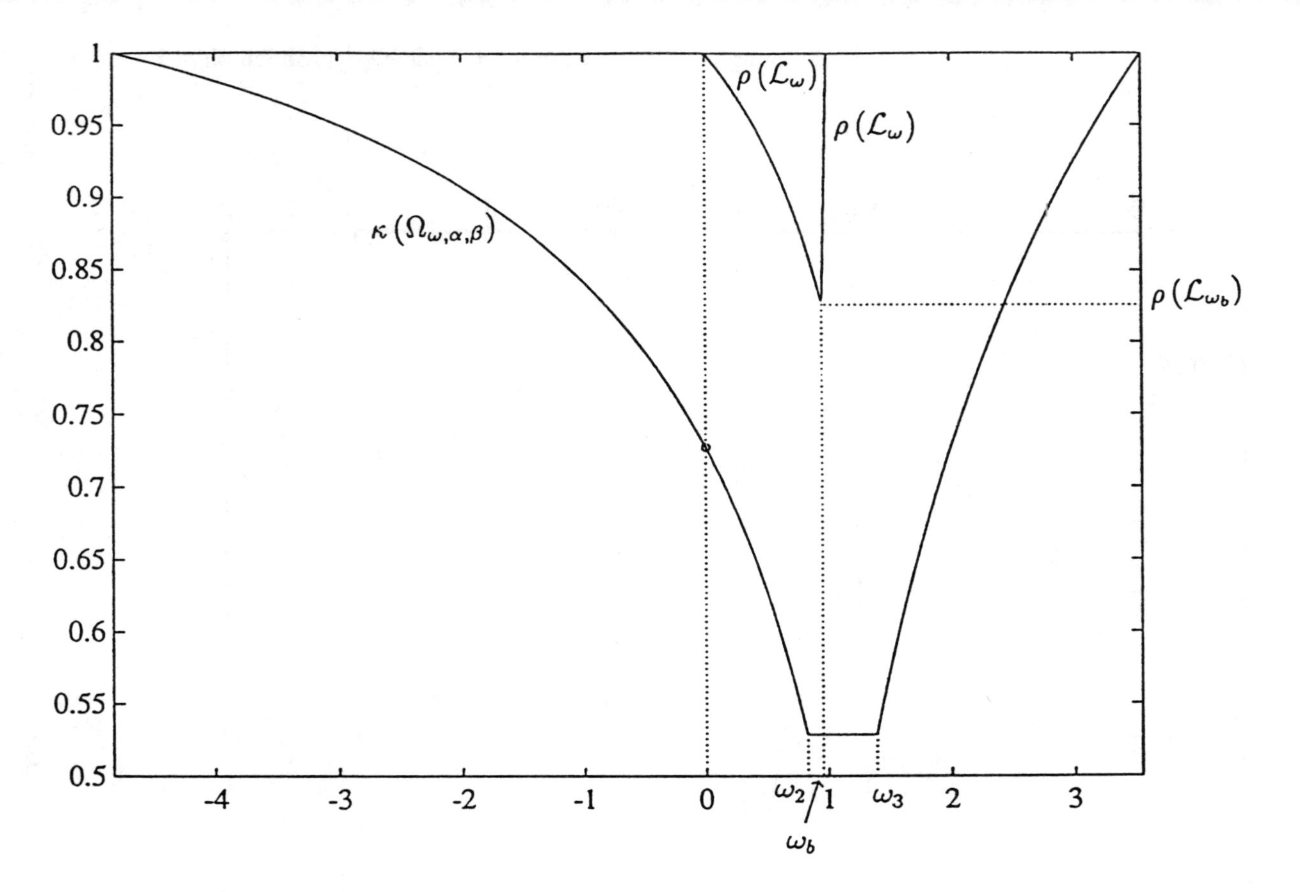

Figure 2: The mixed case: $\rho(\mathcal{L}_\omega)$ and $\kappa(\Omega_{\omega,\alpha,\beta})$ as functions of $\omega \in [\omega_1, \omega_4]$ for $\alpha = 1$ and $\beta = 0.9$.

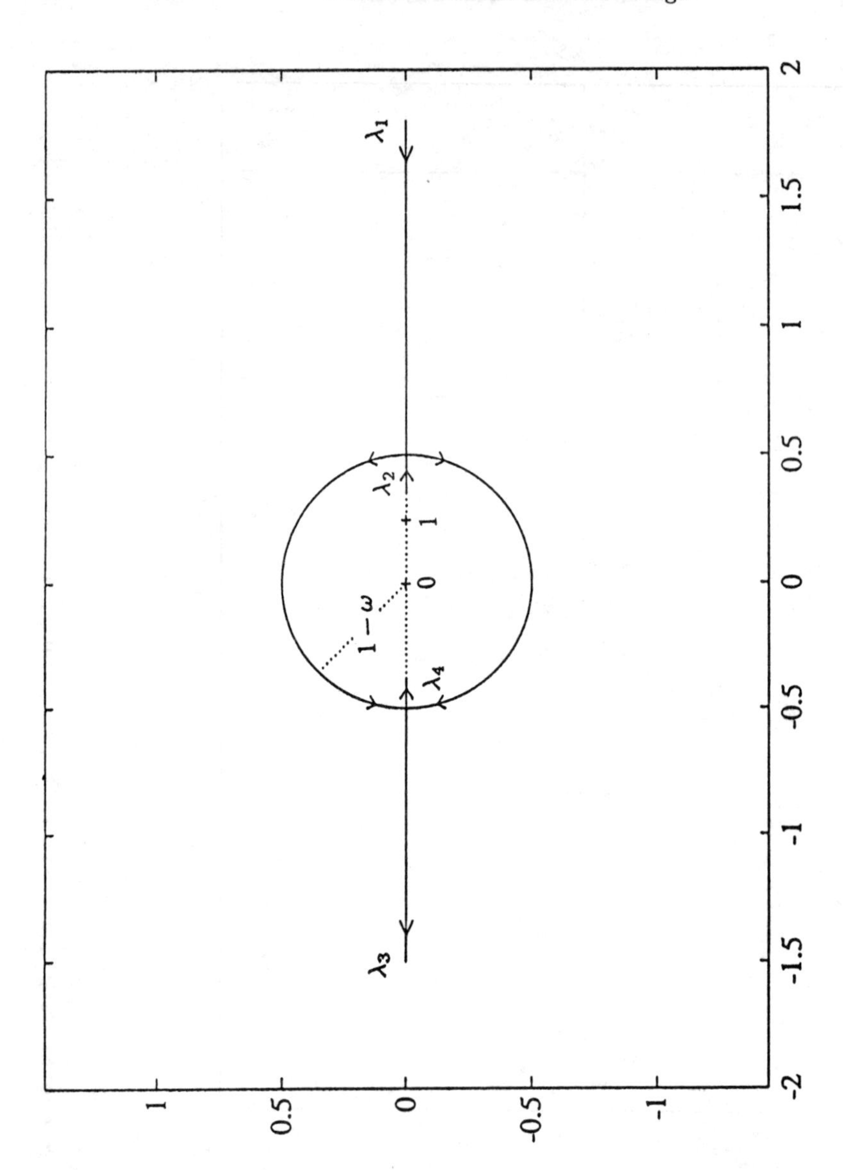

Figure 3a: $\Omega_{\omega,\alpha,\beta}$ for Case 1.

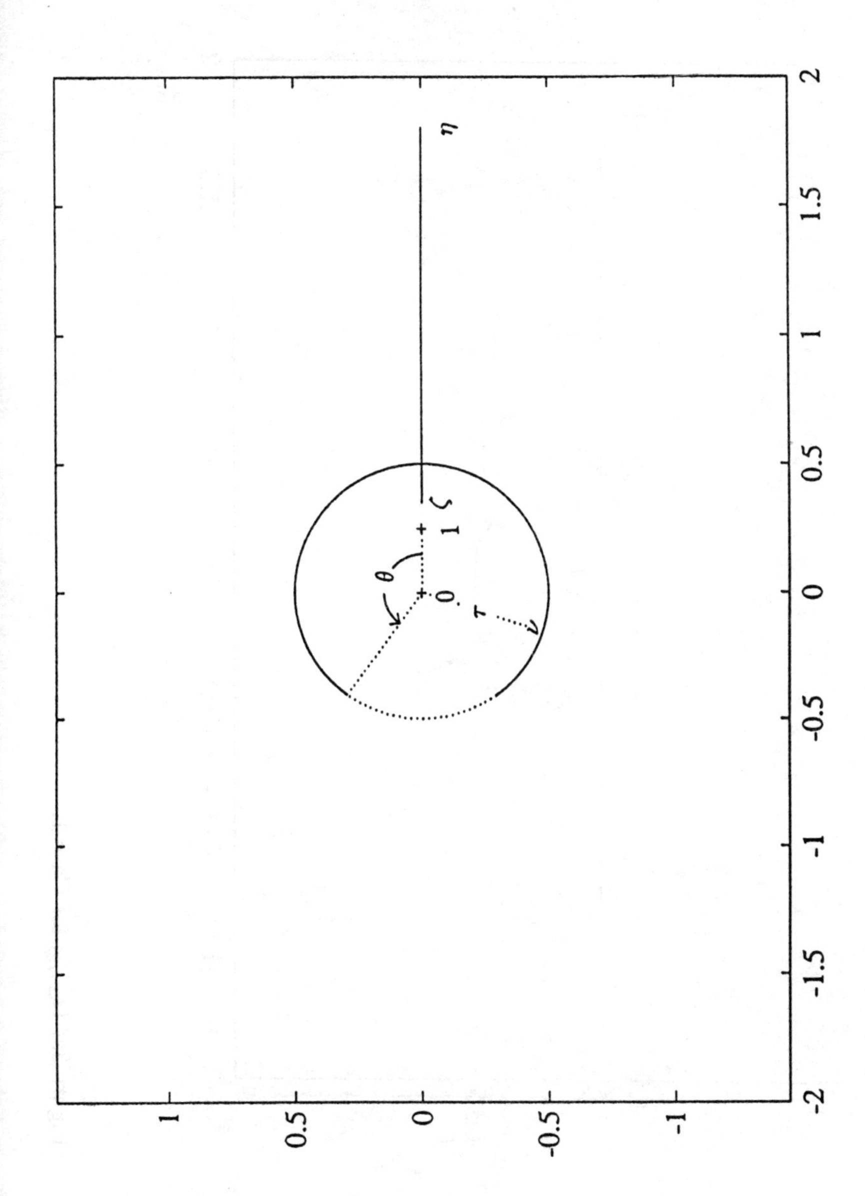

Figure 3b: $D_{\tau,\theta,\zeta,\eta}$.

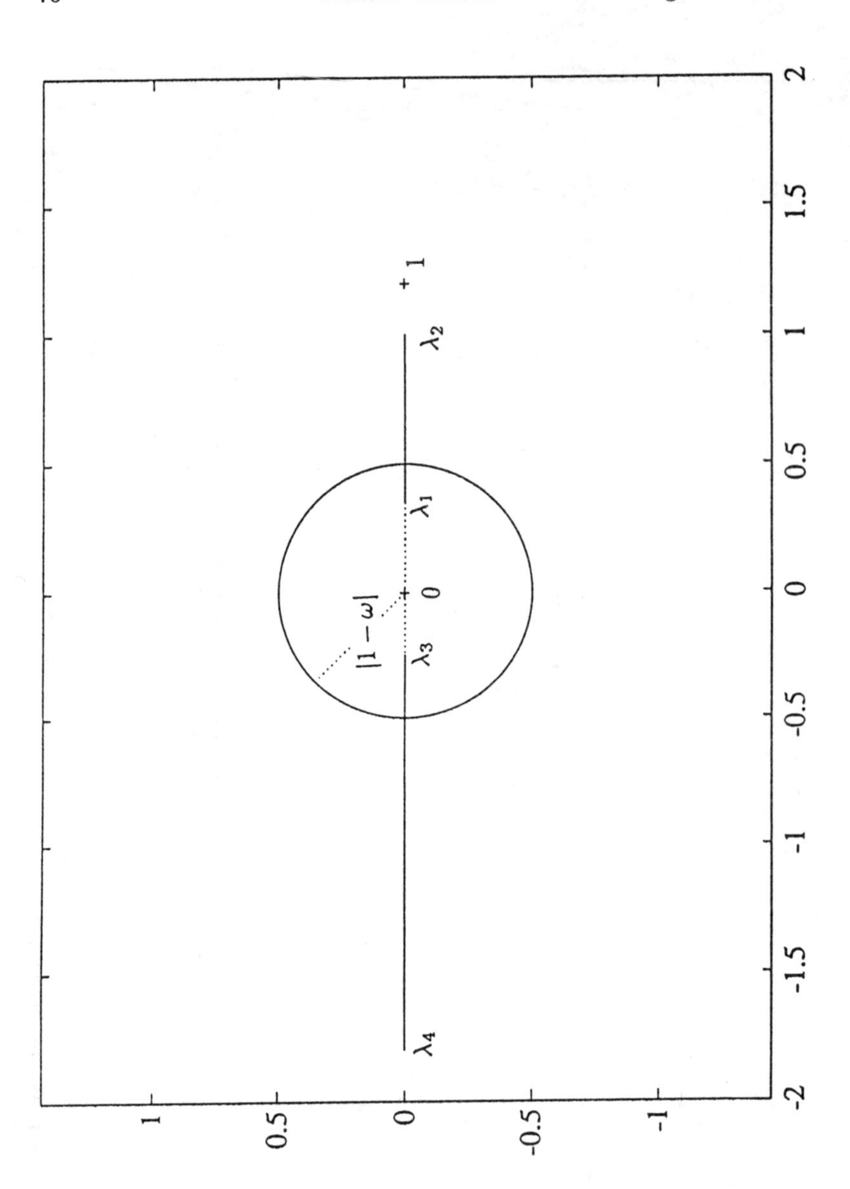

Figure 3c: $\Omega_{\omega,\alpha,\beta}$ for Case 3.

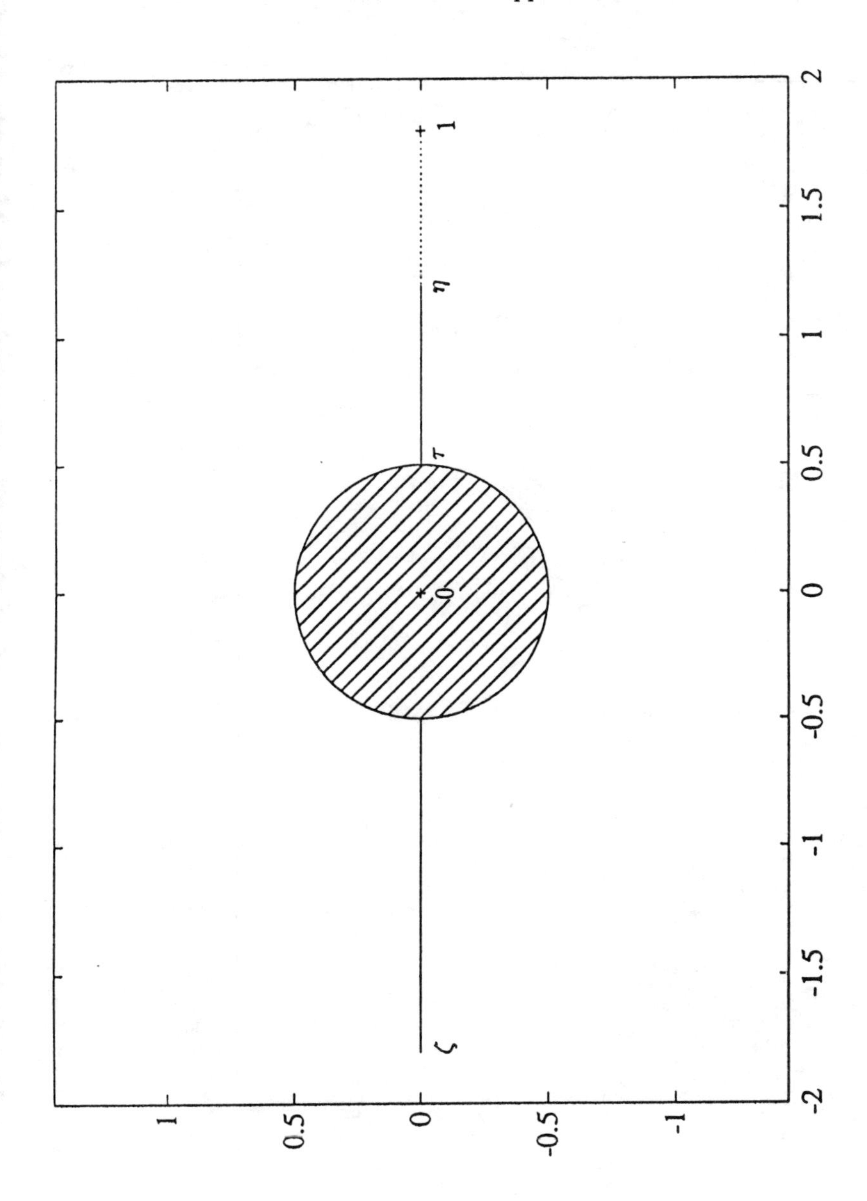

Figure 3d: $B_{\tau,\zeta,\eta}$.

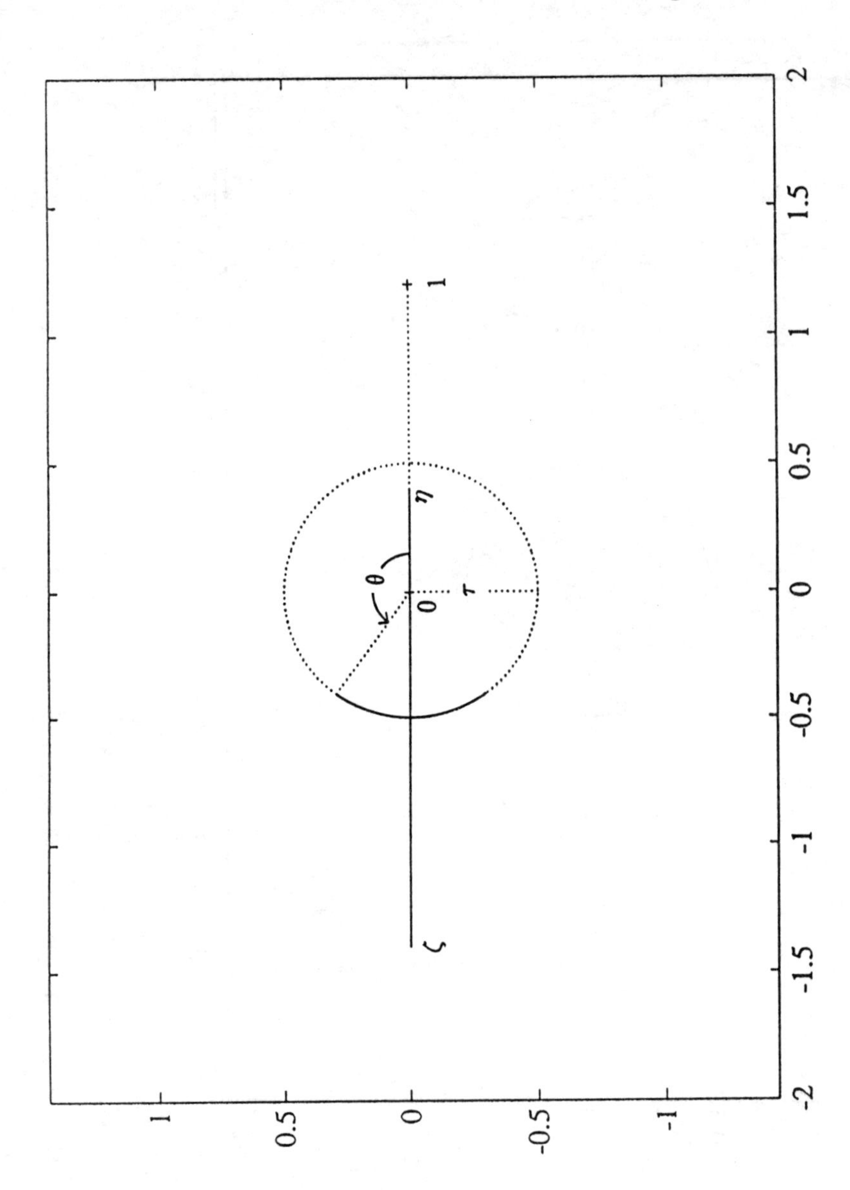

Figure 3e: $C_{\tau,\theta,\zeta,\eta}$.

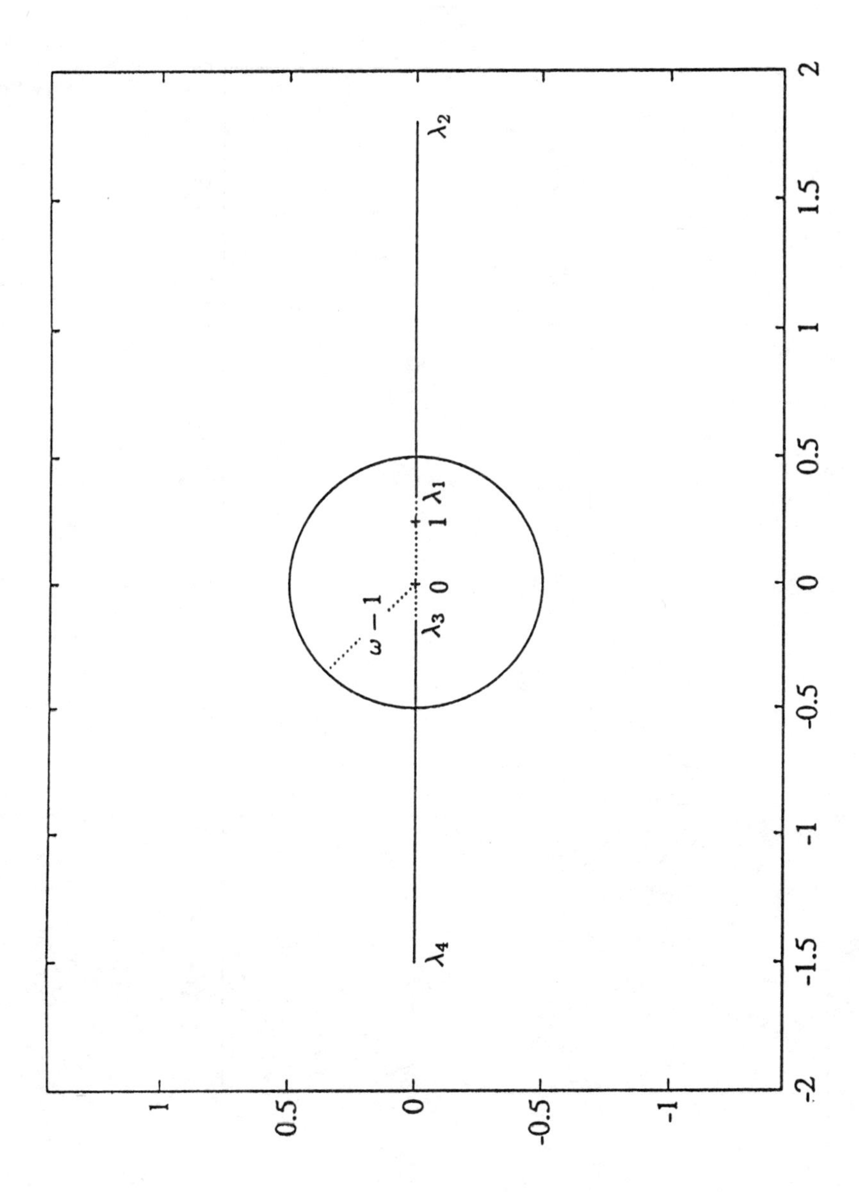

Figure 3f: $\Omega_{\omega,\alpha,\beta}$ for Case 5.

Differential qd algorithms

K. Vince Fernando[*†‡] *and Beresford N. Parlett*[§¶]

Abstract. We have discovered a new implementation of the qd algorithm that has a far wider domain of stability than Rutishauser's version. Our algorithm was developed from an examination of the LR-Cholesky transformation and can be adapted to parallel computation in stark contrast to traditional qd. Our algorithm also yields useful a posteriori upper and lower bounds on the smallest singular value of a bidiagonal matrix.

The zero-shift bidiagonal QR of Demmel and Kahan computes the smallest singular values to maximal relative accuracy and the others to maximal absolute accuracy with little or no degradation in efficiency when compared with the LINPACK code. Our algorithm obtains maximal relative accuracy for all the singular values and runs at least four times faster than the LINPACK code.

Key words. qd, LR algorithm, Cholesky decomposition, singular values, SVD, bidiagonal matrices.

AMS (MOS) subject classification. 65F05.

1 Introduction and Summary

In September 1991 J. W. Demmel and W. M. Kahan were awarded the second SIAM prize in numerical linear algebra for their paper 'Accurate Singular Values of Bidiagonal Matrices' [10], referred to as DK hereafter. Among several valuable results was the observation that the standard bidiagonal QR algorithm used in LINPACK [11], and in many other SVD programs, can be simplified when the shift is zero and, of greater importance, no subtractions occur. The last feature permits very small singular values to be found with (almost) all the accuracy permitted by the data and at no extra cost.

In this paper we show that the DK zero shift algorithm can be further simplified and this simplicity has several benefits. One is that a new algorithm can be

[*]NAG Ltd, Wilkinson House, Jordan Hill, Oxford OX2 8DR, UK and Division of Computer Science, University of California, Berkeley, CA 94708, USA.

[†]Research supported in part by NSF, under grant ASC-9005933

[‡]Research supported in part by ONR, contract N000014-90-J-1372

[§]Department of Mathematics, University of California, Berkeley, CA 94720,USA.

[¶]Supported by ONR, contract N000014-90-J-1372

implemented in either parallel or pipelined format as an $\mathcal{O}(\log_2 n)$ algorithm. This is pursued in a companion paper [13].

Our investigations began with the modest goal of showing that it was preferable to replace the DK zero-shift QR transform by two steps of zero-shift LR implemented in a qd (quotient-difference) format. Root-free algorithms run considerably faster than standard ones. The surprise here is that to keep the high relative accuracy property it is necessary to use a little known variant of qd (the differential form of the progressive qd algorithm or **dqd** [6], [5]). The standard qd will not suffice as we show in Section 4. There are no subtractions in **dqd**. We suspect that Rutishauser discovered **dqd** in 1968, just two years before his death, and we say more about its history in Section 4.

What we want to stress here is that, for reasons we may never know, Rutishauser did not consider the shifted version of **dqd**. Incidentally this differential qd is not to be confused with the continuous analogue of qd (see [2]) and more recent work on QR flows. The trouble with the shifted version of the ordinary qd algorithm is that it cannot recover from a shift that is too large. Consequently qd algorithms have been shackled with very conservative shift strategies, such as Newton's method, and earned the reputation of being slow compared to the QR algorithm. Had Rutishauser considered shifts with differential qd (**dqds** hereafter) he would have realized, as we soon did, that the transformation may be split into two parts. The parts depend on whether the machine is of sequential or parallel type but, in each case, a shift that is too big reveals itself before the old matrix is overwritten and so need not be invoked. An unused shift is not wasted because it gives an improved upper bound on the smallest singular value at a cost less than one qd transformation as well as contributing to an improved shift.

Our approach frees the algorithm to exploit powerful shift strategies while preserving high relative accuracy all the time. In contrast the QR algorithm delivers high relative accuracy only with a zero shift.

Even though our algorithms must find the singular values in order we can use shift strategies that are at least quadratically convergent. This is better than fourth order convergence for QR. When only the smallest few singular values are needed this ordering constraint is a great advantage. Another rather subtle feature is that it is not necessary to make an extra $\mathcal{O}(n)$ check for splitting of the matrix into a direct sum. The necessary information is provided by the auxiliary quantities.

In June 1992 we discovered that our **dqds** algorithm enjoys high relative stability for all shifts provided that they avoid underflow, overflow or divide by zero. Consequently it can be used in a variety of applications (eigenvalues of symmetric or unsymmetric tridiagonals, zeros of polynomials, poles and zeros of transfer functions and many applications involving continued fractions) where Rutishauser's **qd** has been abandoned because of its instability in the general case.

Our error bounds for singular values are significantly smaller than those in DK and the approach is quite transparent. It was this analysis that showed us the possibility of violating positivity while still maintaining maximal relative accuracy for all singular values, not just the small ones.

It gradually dawned on us as we developed the algorithm that we were breaking away from the *orthogonal paradigm* that has dominated the field of matrix computations since the 1960's. It seems to be sacrilegious to be achieving greater accuracy and on average, a six fold speed-up*by simply abandoning QR for something equivalent to LR. See Section 8.2 for details. High accuracy comes from the fact that **dqds** spends most of its time transforming lower triangular 2×2s into upper triangular 2×2s by premultiplication.

Section 3 presents the unifying general result which shows that it is possible to implement the LR-Cholesky algorithm of Rutishauser [3], [7] using orthogonal transformations only. Perhaps this is the key idea exploited in the paper. Since the term LR-Cholesky over describes the algorithm we simply refer to it as the Cholesky Algorithm. Our orthogonal Cholesky algorithm is applicable to dense matrices; this more general case is studied elsewhere [12].

We want to point out the unusual historical lineage of this algorithm. The qd algorithm begat the LR algorithm which then gave rise to the QR algorithm of Francis. This in turn led to the Golub-Kahan and Golub-Reinsch algorithms for singular values of bidiagonal matrices which lead to the DK zero-shift variant. This inspired our orthogonal algorithm of which differential qd is the root-free version. We are back to qd again but with a new implementation.

An exposition of these ideas that contains more details has been submitted to Numerische Mathematik.

2 Notation and Normalization

This paper does not involve vectors very much and so we do not follow Householder conventions. However capital roman letters denote matrices while lower case Roman and Greek letters denote scalars. On the rare occasions when a vector is needed it is denoted by a lower case roman letter in boldface.

As usual the singular values of an $n \times n$ matrix C are arranged in monotone decreasing order and denoted by $\sigma_1, \sigma_2, \ldots, \sigma_n$, their union is $\sigma[C]$.

- We make reference to the QR factorization of a matrix. This is the matrix form of the Gram-Schmidt orthonormalizing process applied to the columns of the matrix in natural order. By convention the diagonal entries of the upper triangular factor R are taken nonnegative. See Golub and Van Loan [14] for details.

- We make reference to the Cholesky factorization of a positive definite matrix into the product of a lower triangular matrix and its conjugate transpose. The factors are unique.

*All our computations are performed on a DECstation 5000/120 using double precision arithmetic (53-bit mantissa).

- We make references to the LR and QR algorithms. These are defined in the appropriate places.

We shall be concerned mainly with bidiagonal matrices which we call B and take them to be *upper* bidiagonal. To save space we write the bidiagonal matrix

$$B = \begin{bmatrix} a_1 & b_1 & & & \\ & a_2 & b_2 & & \\ & & \cdot & \cdot & \\ & & & \cdot & \cdot \\ & & & a_{n-1} & b_{n-1} \\ & & & & a_n \end{bmatrix}$$

as

$$B = bidiag\left\{ \begin{array}{ccccccccccc} & b_1 & & b_2 & & \cdot & & b_{n-2} & & b_{n-1} & \\ a_1 & & a_2 & & \cdot & & \cdot & & a_{n-1} & & a_n \end{array} \right\}.$$

2.1 Normalization

Consider now the effect of a zero value among the parameters of $n \times n$ bidiagonal B.

2.1.1 Superdiagonal. Suppose that $b_k = 0$, $k < n$. Then B may be written as a direct (or diagonal) sum of two bidiagonals B_1 and B_2. Moreover

$$\sigma[B] = \sigma[B_1] \cup \sigma[B_2].$$

This case makes the calculation of singular values easier. Even more important is the fact that our algorithms do not suffer from the failure to detect such a split when it occurs. However, the transition from a linearly convergent shift to a quadratic shift will not occur if the split lies undetected for too long.

2.1.2 Diagonal. Let $a_k = 0$, $k < n$. Since $|\det B| = \prod_{i=1}^{n} \mid a_i \mid = \prod_{i=1}^{n} \sigma_i$ it follows that $\sigma_n = 0$. However some work is needed in order to *deflate* this value, i.e. to find a new B of order $n-1$ yielding the remaining singular values of B. In exact arithmetic one iteration of any of the unshifted algorithms given later is guaranteed to produce the desired B and so this case does not need special treatment. The zero diagonal entry may be driven to the closest end of the matrix.

If $a_k = 0$, $k < n$, at one step of our algorithm and if $a_n = 0$ at the next step then b_{k-1} will also vanish and so produce a split into two bidiagonals.

2.1.3 Signs. If the matrix is real, then using pre and post multiplication by matrices of the form $diag\{\pm 1\}$ any sign pattern may be imposed on the entries of B without changing the singular values. If the matrix is complex, then it could be transformed to a real matrix by pre and post multiplication by matrices of the form $diag\{\exp(i\omega)\}$ where $i^2 = -1$ and ω is real.

There is little loss of generality in assuming, when necessary, that B is of real positive type; all its parameters exceed 0. However in Section 5.3 we address the practical question of when to relax the requirement of positivity.

3 Orthogonal Form of the Cholesky Algorithm

The result given in the theorem below is implicit in proofs that one step of the QR algorithm is equal to two steps of the Cholesky algorithm. Nevertheless it appears not to have been stated explicitly before and was not known to several experts whom we consulted. So for the next few paragraphs we consider full complex matrices. Recall that the Cholesky factorization of a positive definite Hermitian matrix $A(= A^*)$ may be written as $A = LL^*$ where L is lower triangular.

Definition 3.1. *The Cholesky* TRANSFORM *of $A = LL^*$ is*

$$\hat{A} := L^*L.$$

The Cholesky *algorithm*, consisting of successive applications of the Cholesky transformation, is a special case of the LR algorithm.

We now consider the relation between L and $\hat{L}$, the Cholesky factors of A and $\hat{A}$, respectively.

Theorem 3.1. *Let $\hat{A} = \hat{L}\hat{L}^*$ be the Cholesky factorization of the Cholesky transform of positive definite $A = LL^*$. Then*

$$L = Q\hat{L}^*$$

is the QR factorization of L.

Some may prefer the formulation

$$R^* = Q\hat{R}$$

with $A = R^*R$ and $\hat{A} = \hat{R}^*\hat{R}$.

Proof. Since A is positive definite all factors mentioned below are unique. By definition of $\hat{L}$

$$L^*L = \hat{L}\hat{L}^*.$$

We seek invertible F such that

$$L = F\hat{L}^*, \tag{1}$$

$$L^* = \hat{L}F^{-1}. \tag{2}$$

Transpose and conjugate (1) and use invertibility of $\hat{L}$ in (2) to find

$$F^* = \hat{L}^{-1}L^* = F^{-1}.$$

So F is unitary and since $\hat{L}^*$ is upper triangular with positive diagonal Equation (1) above gives the QR factorization of L, as claimed. □

The theorem shows that $\hat{L}$ may be obtained from L by orthogonal transformations without forming $\hat{A}$. Moreover just as QR may be performed with column pivoting so can we obtain the Cholesky factor of a permutation of $\hat{A}$. A general application of Theorem 1 is presented in Fernando and Parlett [12] but here we return to the bidiagonal case.

The basic equation $\hat{L}\hat{L}^t = L^tL$ guarantees that the Cholesky algorithm preserves bandwidth. In particular, bidiagonal B gives rise to tridiagonal $A = B^tB$ and a bidiagonal $\hat{B}$. In order to study how $\hat{B}$ is derived from B, let

$$B = bidiag\left\{\begin{matrix} & b_1 & & b_2 & & . & & b_{n-2} & & b_{n-1} & \\ a_1 & & a_2 & & . & & . & & a_{n-1} & & a_n \end{matrix}\right\}.$$

$$\hat{B} = bidiag\left\{\begin{matrix} & \hat{b}_1 & & \hat{b}_2 & & . & & \hat{b}_{n-2} & & \hat{b}_{n-1} & \\ \hat{a}_1 & & \hat{a}_2 & & . & & . & & \hat{a}_{n-1} & & \hat{a}_n \end{matrix}\right\}.$$

where $\hat{B}^t\hat{B} = BB^t$. By Theorem 1

$$B^t = Q\hat{B}.$$

The matrix Q may be written as a product of $(n-1)$ plane rotation matrices [14],

$$Q = G_1G_2\ldots G_{n-1}.$$

Before the annihilation of the subdiagonal element b_k, the active part of the matrix is of the form,

$$\begin{matrix} 0 & \hat{a}_{k-1} & \hat{b}_{k-1} & & \\ & 0 & \tilde{a}_k & 0 & \\ & & b_k & a_{k+1} & 0 \\ & & & b_{k+1} & a_{k+2} \end{matrix} \tag{3}$$

and after the plane rotation G_k^t, the matrix becomes

$$\begin{matrix} 0 & \hat{a}_{k-1} & \hat{b}_{k-1} & & \\ & 0 & \hat{a}_k & \hat{b}_k & \\ & & 0 & \tilde{a}_{k+1} & 0 \\ & & & b_{k+1} & a_{k+2} \end{matrix} \tag{4}$$

Formally we may set $B^{(0)} = B^t$ and, for $k = 1, \ldots, n-1$

$$B^{(k)} = G_k^t B^{(k-1)}. \tag{5}$$

Finally $\hat{B} = B^{(n-1)}$ and, from (3) and (4), with $\tilde{a}_1 = a_1$ and $c_k^2 + s_k^2 = 1$,

$$\hat{a}_k = \sqrt{\tilde{a}_k^2 + b_k^2} = \tilde{a}_k / c_k \tag{6}$$

$$s_k = b_k / \hat{a}_k$$

$$c_k = \tilde{a}_k / \hat{a}_k \tag{7}$$

$$\hat{b}_k = s_k a_{k+1} = b_k a_{k+1} / \hat{a}_k \tag{8}$$

$$\tilde{a}_{k+1} = c_k a_{k+1} = \tilde{a}_k a_{k+1} / \hat{a}_k.$$

There is some redundancy in the equations given above but their most important property is the absence of subtractions. This ensures high relative accuracy in the new entries $\hat{a}_i$ and $\hat{b}_i$. Observe that neither s_k nor c_k is needed explicitly to compute the new entries. To the best of our knowledge the algorithm given below is new. For reasons that appear in the next section we call it the *Orthogonal qd-Algorithm* or **oqd**. It is convenient to use

$$cabs(x,y) = \sqrt{x^2 + y^2} \tag{9}$$

whose name stands for the *c*omplex *abs*olute value of $x + iy$. In numerical computing (e.g. Eispack), an alternative name for *cabs* is *pythag*.

Algorithm 1 (oqd)

$$
\begin{aligned}
&\tilde{a} := a_1 \\
&\mathit{for}\, k = 1, n-1 \\
&\quad \hat{a}_k := cabs(\tilde{a}, b_k) \\
&\quad \hat{b}_k := b_k * (a_{k+1}/\hat{a}_k) \\
&\quad \tilde{a} := \tilde{a} * (a_{k+1}/\hat{a}_k) \\
&\mathit{end\, for} \\
&\hat{a}_n := \tilde{a}
\end{aligned}
$$

This algorithm will undergo several transformations in the following pages before we are ready to implement it. Nevertheless, even at this stage, two applications of it are slightly better (fewer multiplications) than the DK Zero Shift QR algorithm [10] described briefly in our Section 10.

The inner loop comparisons given in Table 1 are based on one QR step which is equal to two LR steps. We have taken into account the common sub-expression $a_{k+1}/\hat{a}_k$ in the estimation of the complexity of **oqd** (Algorithm 1).

	DK	oqd
cabs	2	1*2
divisions	2	1*2
multiplications	6	2*2
conditionals	1	0
assignments	7	3*2
auxiliary variables	6	1

Table 1: Complexity of Demmel-Kahan and **oqd**

DK uses six auxiliary variables while **oqd** needs only one. The memory traffic is essentially determined by the number of variables, arithmetic operations and assignment statements. In most advanced architectures, memory access is more expensive than floating-point operations and in such machines the **oqd** will be very advantageous because of fewer read and write operations.

4 The Quotient Difference Algorithm

It is easy to avoid taking the square roots that appear in **oqd** (Algorithm 1) . Define $b_n := 0$ and $q_k = a_k^2$, $e_k = b_k^2$, $k = 1, 2, \ldots, n$. By simply squaring each assignment in **oqd** (Algorithm 1) one obtains an algorithm that turns out to be a little known variant of the quotient difference algorithm. Rutishauser developed his qd algorithm in several papers from 1953 or 1954 (e.g. [1]) until his early death in 1970 but this variant appeared in English only in 1990 in [6] which is the translation of the German original [5] published in 1976. The full list of the papers on qd by Rutishauser can be found in the above mentioned books which were published posthumously.

In the notes at the end of [1] and at the end of volume 2 of [5] this variant is called the *differential form* of the progressive qd algorithm or **dqd**. These notes were based on unfinished manuscripts of Rutishauser.

Algorithm 2 (dqd)

$$
\begin{aligned}
& d := q_1 \\
& \textit{for } k := 1, n-1 \\
& \quad \hat{q}_k := d + e_k \\
& \quad \hat{e}_k := e_k * (q_{k+1}/\hat{q}_k) \\
& \quad d := d * (q_{k+1}/\hat{q}_k) \\
& \textit{end for} \\
& \hat{q}_n := d
\end{aligned}
$$

The implementation of **dqd** (Algorithm 2) requires only 1 division, 2 multiplies, and 1 addition in the inner loop. No subtractions occur.

The intermediate variable d may be removed. At step k, $d = d_k$ and the trick is to write it as a difference.

$$d_{k+1} = c_k^2 q_{k+1} = q_{k+1} - s_k^2 q_{k+1} = q_{k+1} - \hat{e}_k .$$

Algorithm 3 (qd)

$$
\begin{aligned}
& \hat{e}_0 = 0 \\
& \textit{for } k := 1, n-1 \\
& \quad \hat{q}_k := (q_k - \hat{e}_{k-1}) + e_k \\
& \quad \hat{e}_k := e_k * q_{k+1}/\hat{q}_k \\
& \textit{end for} \\
& \hat{q}_n := q_n - \hat{e}_{n-1}
\end{aligned}
$$

Table 2 compares the complexity of orthogonal, differential and standard qd algorithms.

We hasten to add that Rutishauser did not derive the qd algorithm from our Theorem 1 but from ideas described in Section 11.

	oqd	dqd	qd
cabs	1	0	0
divisions	2	1	1
multiplications	4	2	1
additions	1	1	1
subtractions	0	0	1
assignments	3	3	2
auxiliary variables	1	1	0

Table 2: Complexity of **oqd**, **dqd** and **qd**

For positive B, **dqd** and **qd** are stable in the sense that all intermediate quantities are bounded by $\|B\|^2$. Singular value errors provoked by finite precision arithmetic will be tiny compared to σ_1^2. This is satisfactory for many purposes and it was not generally appreciated until the DK paper appeared that bidiagonal matrices do determine all their singular values, however small, to the same relative precision enjoyed by the matrix entries. Since such accuracy can be achieved for little extra cost it seems only right to do so. These considerations lead us to abandon **qd** and concentrate on **dqd** and **oqd**.

Example 1 Here is a bidiagonal Toeplitz matrix with $a_i = 1$, $b_i = 256$ ($q_i = 1$, $e_i = 65536$) for all i. The results of our **dqd** algorithm are given in Table 3. Note that $\sqrt{q_{64}} = 1.9093060930437717 \times 10^{-152} \approx 2^{-504}$ gives σ_{64} correct to full machine precision.

The results for **qd** were identical to **dqd** except that the crucial element q_{64} became zero in both steps. Hence **qd** is not suitable for computation of small singular values with high relative accuracy.

	after the first pass	after the second pass
q_1	6.5537000000000000D+04	6.5537999984741444D+04
q_2	6.5536000015258556D+04	6.5536000061032595D+04
q_3	6.5536000000000233D+04	6.5536000000001397D+04
q_4 to q_{63}	6.5536000000000000D+04	6.5536000000000000D+04
q_{64}	3.6455053829317361D-304	3.6454497569340717D-304
e_1	9.9998474144376459D-01	9.9995422572819948D-01
e_2	9.9999999976717291D-01	9.9999999883589297D-01
e_3	9.9999999999999645D-01	9.9999999999997513D-01
e_4 to e_{62}	1.0000000000000000D+00	1.0000000000000000D+00
e_{63}	1.0000000000000000D+00	5.5625997664363648D-309

Table 3: Numerical results for Example 1

Example 2 We have rerun Example 1 but with a smaller value of ($n = 5$) and the results are given in Table 4. For this example, $\sigma_5 = \sqrt{q_5} = 2.3282709094019085 \times 10^{-10}$ which is correct to full machine precision. For comparison, the answer given by the LINPACK SVD routine **dsvdc** (which is based on the Golub-Reinsch algorithm) is $2.3282704794711363 \times 10^{-10}$ which gets 7 of the 15 digits correct.

Using **qd** we got almost identical results except that q_5 is zero in both sweeps. Thus, σ_5 is zero according to the **qd** algorithm. Thus, **qd** does not deliver as much accuracy as Golub-Reinsch; in fact it can be shown that **qd** sometimes delivers zero for singular values as large as $\sqrt{macheps} * \|B\|$.

	after the first pass	after the second pass
q_1	6.5537000000000000D+04	6.5537999984741449D+04
q_2	6.5536000015258551D+04	6.5536000061032593D+04
q_3	6.5536000000000238D+04	6.5536000000001395D+04
q_4	6.5536000000000000D+04	6.5536000000000000D+04
q_5	5.4209281443662679D-20	5.4208454275671899D-20
e_1	9.9998474144376457D-01	9.9995422572819948D-01
e_2	9.9999999976717293D-01	9.9999999883589292D-01
e_3	9.9999999999999642D-01	9.9999999999997509D-01
e_4	1.0000000000000000D+00	8.2716799077854419D-25

Table 4: Numerical results for Example 2

Some people do not like root free algorithms (e.g. **dqd**) because they limit the domain of the matrices to which they can be applied. For example, a bidiagonal B whose singular values vary from 10^{30} to 10^{-30} could be diagonalized in single precision on an IBM machine by **oqd** (Algorithm 1) but not by **dqd** (Algorithm 2) because of overflow and underflow.

We conclude this section by pointing out that **qd** (Algorithm 3), the standard qd algorithm, consists of the so-called *rhombus rules* arranged in computational form and these rules are a direct consequence of the defining equation

$$BB^t = \hat{B}^t\hat{B}.$$

Equate the (k, k) entry on each side to obtain

$$a_k^2 + b_k^2 = \hat{b}_{k-1}^2 + \hat{a}_k^2 \tag{10}$$

$$q_k + e_k = \hat{e}_{k-1} + \hat{q}_k.$$

and equate the $(k, k+1)$ entry on each side to obtain

$$b_k a_{k+1} = \hat{a}_k \hat{b}_k \tag{11}$$

$$e_k q_{k+1} = \hat{q}_k \hat{e}_k.$$

The rhombus rules can be also derived from $B^t = Q\hat{B}$ by noting that orthogonal transformation Q changes neither the norms nor the inner products of the columns. The reason for the name rhombus rule is indicated in Figure 3 of Section 11.

5 Incorporation of Shifts

Rutishauser introduced shifts into the **qd** almost from the beginning and we could simply quote him. Unfortunately he does not give any explanation of how he derived the appropriate modification of **qd** (given in Section 4). So we provide one at the end of Section 5.1.

5.1 Shifted qd Algorithms

In eigenvalue calculations, shifts are natural and can be easily incorporated since

$$\lambda(A - \tau^2 I) = \lambda(A) - \tau^2$$

where τ^2 is the shift and $\lambda(A)$ indicates an eigenvalue of A. Thus, by subtracting τ^2 from the diagonals of the matrix, we can introduce origin shifts into the Cholesky algorithm.

A shift τ can be introduced into **oqd** (Algorithm 1, Section 3) by modifying statements involving $\hat{a}$ and $\tilde{a}$.

Algorithm 4 (oqds)

$$\begin{aligned}
&\tilde{a} := a_1 \\
&\textit{for } k = 1, n-1 \\
&\quad \hat{a}_k := \sqrt{\tilde{a}^2 + b_k^2 - \tau^2} \\
&\quad \hat{b}_k := b_k * (a_{k+1}/\hat{a}_k) \\
&\quad \tilde{a} := \sqrt{\tilde{a}^2 - \tau^2} * (a_{k+1}/\hat{a}_k) \\
&\textit{end for} \\
&\hat{a}_n := \sqrt{\tilde{a}^2 - \tau^2}
\end{aligned}$$

It may be verified that $\hat{B}^t \hat{B} = BB^t - \tau^2 I$. To keep $\hat{B}$ real the shift must satisfy

$$\tau \leq \sigma_n[B] \tag{12}$$

but this constraint is not formally necessary for **dqd** (Algorithm 2) which uses

$$\hat{q}_k := d_k + e_k - \tau^2.$$

Algorithm 5 (dqds)

$$\begin{aligned}
&d := q_1 - \tau^2 \\
&\textit{for } k := 1, n-1 \\
&\quad \hat{q}_k := d + e_k \\
&\quad \hat{e}_k := e_k * (q_{k+1}/\hat{q}_k) \\
&\quad d := d * (q_{k+1}/\hat{q}_k) - \tau^2 \\
&\textit{end for} \\
&\hat{q}_n := d
\end{aligned}$$

The constraint (12) is also unnecessary for **qd**.

Algorithm 6 (qds)

$$
\begin{aligned}
&\hat{e}_0 = 0\\
&\textit{for } k := 1, n-1\\
&\quad \hat{q}_k := (q_k - \hat{e}_{k-1}) + e_k - \tau^2\\
&\quad \hat{e}_k := e_k * q_{k+1}/\hat{q}_k\\
&\textit{end for}\\
&\hat{q}_n := q_n - \hat{e}_{n-1} - \tau^2
\end{aligned}
$$

All that is lacking is an analogue of the orthogonal connection (Theorem 1)

$$B^t = Q\hat{B}.$$

For that it is necessary to abandon square matrices and write

$$\begin{bmatrix} B^t \\ O \end{bmatrix} = Q \begin{bmatrix} \hat{B} \\ \tau I \end{bmatrix}.$$

The new Q is $2n \times 2n$ and is not unique. However its first n rows are uniquely determined by B and τ for $\tau \leq \sigma_n[B]$.

It is at this point that the superiority of the qd formulation becomes clear. DK showed that the standard Golub-Reinsch bidiagonal QR algorithm may be simplified when the shift is zero; see Section 10 for the details. Our algorithms (1,2, or 3) are already simpler than the DK zero shift QR and they also permit use of a non-zero shift with no impediment to pipelined or parallel implementation or high relative accuracy. See [13] for details. This is strong evidence that our formulation is the natural one.

5.2 The Two Phase Implementation

At first sight the auxiliary quantities d_i, $i = 1, \ldots, n$ that occur in **dqd** are seen as the price to be paid for securing high relative accuracy. On further consideration they may be seen as an attractive feature that permits an aggressive shift strategy that also preserves high relative accuracy in the computed singular values. Moreover, as an extra bonus, we find that the vector $\mathbf{d} = (d_1, \ldots, d_n)$ may be computed in $\mathcal{O}(\log_2 n)$ steps in a parallel computer using the technique called parallel prefix operation in computer science writings, see [9].

Consider next the implementation of **dqds**. The auxiliary quantities d_i may be computed prior to any modification of q and e since

$$
\begin{aligned}
d_{k+1} &= d_k q_{k+1}/\hat{q}_k - \tau^2\\
&= d_k q_{k+1}/(d_k + e_k) - \tau^2.
\end{aligned} \tag{13}
$$

An alternative formulation is

$$d_{k+1} = \frac{q_{k+1}}{1 + e_k/d_k} - \tau^2 \tag{14}$$

but a division costs more than a multiplication.

It is at this point that one sees the advantage of arithmetic units that conform to the ANSI/IEEE floating point standard 754: there is no need to test at each instance of (13) or (14) to prevent division by zero. The occurence of a k with $d_k = \infty$ does no harm. It signals that

$$\sigma_n^2[B] < \tau^2$$

and the transformation of B to $\hat{B}$ (Phase 2) should not be completed. The effort in running (13) is not wasted because it yields a new upper bound on $\sigma_n^2[B]$.

Using (13), $d_k = \infty$ yields $d_{k+1} = \infty/\infty = NaN$ (not a number) and then $\hat{q}_i = NaN$ for $i > k+1$. Using (14), $d_k = \infty$ yields $d_{k+1} = q_{k+1} - \tau^2$ which is a better answer.

5.3 Almost Positive Bidiagonals

The standard qd algorithm is well defined for most shifts but it may not be stable in an absolute sense; i.e. the new array $\{\hat{q}, \hat{e}\}$ may be far greater than old one $\{q, e\}$. Rutishauser proved stability under the assumption of positivity and took great care in his implementation to preserve this property.

Our **dqds** algorithm has the advantage of maintaining relative stability in the positive case and, fortunately, even beyond. We currently impose the requirement

$$\tau^2 < \frac{1}{2}\sigma_{n-1}^2[B_{n-1}] + e_{n-1}$$

where B_{n-1} is the leading principal submatrix of B_n because it ensures that the only entries in $\{\hat{q}, \hat{e}\}$ that could go negative are $\hat{e}_{n-1}$ and $\hat{q}_n$. Our goal is to choose τ (actually τ^2) to make $\hat{q}_n$ as small as possible and hence

$$\tau^2 \approx d_n = q_n(1 - e_{n-1}/\hat{q}_{n-1}).$$

Notice how strongly d_n depends on $sign(e_{n-1})$ and $sign(q_n)$ since $\hat{q}_{n-1}$, though unknown, remains positive. There are four possible configurations in the asymptotic regime ($\tau^2 < \frac{1}{2}d_{n-1}+e_{n-1}$) and we designate them by sign pairs: $(sign(e_{n-1}), sign(q_n))$. Each time that **dqds** is invoked there is no doubt about $sign(\hat{e}_{n-1})$ but $sign(\hat{q}_n)$ will not be predictable since the aim is to have $\hat{q}_n = 0$.

A careful study of the last three assignments in **dqds** shows the following possible paths the iteration could follow. Since we do not expect more than 2 steps before convergence (and deflation) some edges may not be traversed.

If $\tau 2 < \sigma_n$

$$\begin{array}{ccc} (+,+) & \longrightarrow & (+,+) \\ (+,-) & \longrightarrow & (-,+) \\ (-,+) & \longrightarrow & (-,-) \\ (-,-) & \longrightarrow & (+,+) \end{array}$$

If $\tau 2 > \sigma_n$

$$\begin{array}{ccc} (+,+) & \longrightarrow & (+,-) \\ (+,-) & \longrightarrow & (-,-) \\ (-,+) & \longrightarrow & (-,+) \\ (-,-) & \longrightarrow & (+,-) \end{array}$$

6 Bounds for σ_{min}

6.1 A Posteriori Bounds for the Smallest Singular Value

Our **oqd**(Algorithm 1 in Section 3) transforms B to $\hat{B}$ by making use of n auxiliary quantities $\tilde{a}_k, k = 1, n$. It is possible to give a nice interpretation of the $\tilde{a}_k$ that leads to useful bounds on σ_{min}. This result was also obtained by Rutishauser but his treatment was not based on orthogonal rotations although he knew the matrix interpretation of qd.

If we think of the matrix B^t being transformed into $\hat{B}$ one column at a time in $(n-1)$ little steps then at the end of Step $(k-1)$ row k is a singleton. That is the key technical observation. To describe the situation we refer back to Section 3 and let $Q_k = (G_1 G_2 \ldots G_{k-1})^t$ be the product of the first $(k-1)$ plane rotations used in the reduction process. Thus

$$B^{(k)} = Q_k B^t = \begin{bmatrix} \hat{a}_1 & \hat{b}_1 & & & & & & & & & \\ 0 & \hat{a}_2 & \hat{b}_2 & & & & & & & & \\ & 0 & . & . & & & & & & & \\ & & 0 & \hat{a}_{k-1} & \hat{b}_{k-1} & & & & & & \\ & & & 0 & \tilde{a}_k & 0 & & & & & \\ & & & & b_k & a_{k+1} & 0 & & & & \\ & & & & & b_{k+1} & a_{k+2} & & & & \\ & & & & & & . & . & 0 & & \\ & & & & & & & b_{n-2} & a_{n-1} & 0 \\ & & & & & & & & b_{n-1} & a_n \end{bmatrix} \tag{15}$$

Note that $Q_k B^t$ coincides with $\hat{B}$ in rows $1, 2, \ldots, k-1$ and with B^t in rows $k+1, \ldots, n$ while orthogonal Q_k coincides with I_n in rows $k+1, \ldots, n$.

Theorem 6.1 (Bounds for σ_{min} without shifts) *Apply the* **dqd** *transformation to a positive bidiagonal B (see Algorithm 1) to produce $\hat{B}$ and $\tilde{a}_1, \tilde{a}_2, \ldots, \tilde{a}_n$. Then*

1. $\sigma_n \le \min_k\{\tilde{a}_k\}$

2. $[(BB^t)^{-1}]_{k,k} = \tilde{a}_k^{-2}$

3. $(\sum_{k=1}^n \tilde{a}_k^{-1})^{-1} \le (\sum_{k=1}^n \tilde{a}_k^{-2})^{-1/2} \le \sigma_n$.

Proof. Since singular values are invariant under orthogonal transformations and transposition

$$\sigma_n[B] = \sigma_n[Q_k B^t] \leq \|\mathbf{u}_k^t Q_k B^t\| = \tilde{a}_k$$

where $\mathbf{u}_k$ is the kth column of the identity matrix. The kth row of $Q_k B^t$ is a singleton;

$$\mathbf{u}_k^t Q_k B^t = \tilde{a}_k \mathbf{u}_k^t.$$

Transposing and rearranging gives

$$\tilde{a}_k^{-1} Q_k \mathbf{u}_k = B^{-1}\mathbf{u}_k$$

$$\tilde{a}_k^{-2} = (B^{-t}B^{-1})_{k,k}$$

as claimed. Note that

$$\sigma_n^{-2} \leq \sum_{i=1}^{n} \sigma_i^{-2} = \|B^{-1}\|_F^2 = trace[(BB^t)^{-1}] = \sum_{i=1}^{n} \tilde{a}_i^{-2}.$$

Finally we get the required result by considering the one and two norms of the vector $(\tilde{a}_1^{-1}, \tilde{a}_2^{-1}, \ldots \tilde{a}_n^{-1})$. □

We can compute bounds on $\sigma_{min}[B]$ even when the algorithm is used with shifts τ provided that $\tau \leq \sigma_{min}[B]$. Formally the reduction

$$\begin{pmatrix} B^t \\ 0 \end{pmatrix} \rightarrow \begin{pmatrix} \hat{B} \\ \tau I \end{pmatrix}$$

requires $2(n-1)$ plane rotations (not just $n-1$) because the rotation G_i in $(i, i+1)$ must be preceded by a rotation $\vec{G}_i$ in plane $(i, n+i)$ in order to introduce τ into position $(n+i, i)$. Thus the rows $1, \ldots, k-1$ and $n+1, \ldots, n+k-1$ of

$$Q_k \begin{pmatrix} B^t \\ 0 \end{pmatrix} \quad \text{and} \quad \begin{pmatrix} \hat{B} \\ \tau I \end{pmatrix}$$

are coincident. Also the rows $k+1, \ldots, n$ and $n+k, \ldots, 2n$ of

$$Q^t \begin{pmatrix} B^t \\ 0 \end{pmatrix} \quad \text{and} \quad \begin{pmatrix} B^t \\ 0 \end{pmatrix}$$

are the same. However row k is still a singleton in fact

$$u_k^t Q_k \begin{pmatrix} B^t \\ 0 \end{pmatrix} = \tilde{a}_k u_k^t. \tag{16}$$

Theorem 6.2 (Bounds for σ_{min} with shifts) *If the* **dqds** *algorithm with shift τ transforms positive bidiagonal B into positive $\hat{B}$ with auxiliary quantities $\tilde{a}_1, \ldots, \tilde{a}_n$ then*

1. $\sigma_n \leq \min_k\{\tilde{a}_k\}$
2. $[(BB^t)^{-1}]_{k,k} = \tilde{a}_k^{-2}\|x_k\|^2 < \tilde{a}_k^{-2}$.

where, in (16), $u_k^t Q_k := (x_k^t, y_k^t)$, and x and y each have n entries.

Proof. Since singular values are invariant under orthogonal transformation and transposition,

$$\sigma_n[B] = \sigma_n \left[Q_k \begin{pmatrix} B^t \\ 0 \end{pmatrix} \right] \leq \|u_k^t Q_k \begin{pmatrix} B^t \\ 0 \end{pmatrix} \| = \tilde{a}_k.$$

The last equality uses (16). To establish the second result transpose (16) to obtain

$$\begin{bmatrix} B & 0 \end{bmatrix} Q_k^t u_k = Bx_k = u_k \tilde{a}_k.$$

Since B is invertible,

$$\tilde{a}_k^{-1} x_k = B^{-1} u_k,$$

$$\tilde{a}_k^{-2}\|x_k\|^2 = u_k^t B^{-t} B^{-1} u_k = [(BB^t)^{-1}]_{k,k}.$$

□

Remark: Since the $\tilde{a}_k$ are monotone decreasing in τ a successful **dqds** transformation produces a better upper bound and a worse lower bound than does **dqd**. Fortunately it is the upper bound that plays an active role in our implementation.

6.2 The Johnson Bound

For a general complex matrix C, a Gersgorin-type bound for σ_{min} is given by Johnson (see [15]),

$$\sigma_{min} \geq \max\{0, \theta\}$$

where

$$\theta = \min_i \left\{ |c_{i,i}| - \frac{1}{2} \sum_{k \neq i} |c_{k,i}| + |c_{i,k}| \right\}.$$

For a positive bidiagonal B, this simplifies to

$$\theta = \min_i \left\{ a_i - \frac{1}{2}(b_i + b_{i-1}) \right\}$$

and ultimately this becomes

$$\theta = a_n - \frac{1}{2} b_{n-1}.$$

7 Effects of Finite Precision

7.1 Error Analysis - Overview

One of the benefits of the simplicity of our algorithms **oqd** and **dqd** is that their analysis is relatively easy. The DK zero shift QR transformation, though simpler than the Golub/Reinsch transformation, is complicated enough to defy anything but a forward error analysis. After heroic struggles with innumerable details DK establish the error bound quoted in Section 10.4.

When discussing this result and our own analyses it is convenient to use the acronym *ulp* which stands for *u*nits in the *l*ast *p*lace held. It is the natural way to refer to *relative* differences between numbers. When a result is correctly rounded the error is not more than half an *ulp*. In this section we usually omit the ubiquitous phrase 'at most' qualifying errors and modifications.

Our algorithms still do not admit a pure backward error analysis, the computed output $\hat{B}$ is not the exact output from a matrix very close to B. Nevertheless we can use a hybrid interpretation involving both backward and forward interpretation.

Whereas DK's zero shift guarantees that each computed singular value is in error by no more than $69n^2$ *ulps* our **dqds** algorithm causes no more than $4n$ *ulps* change using any properly chosen shift. However the main point is that our analysis is easy to grasp.

The next subsection establishes this strong property of **dqds**. A similar result holds for **oqds** but the square roots and squaring provoke a slightly larger bound.

The trick of the proof is to define $\vec{B}$ so that the computed auxiliary quantities $\{d_i\}$ are exact outputs of **dqds**. The difference between $\breve{B}$ and $\hat{B}$ is the forward error.

At the beginning of the paper we made much of the fact that algorithms **oqd** and **dqd** required no subtractions. Yet, in the interest of efficiency, we have introduced shifts and quietly brought back subtraction. The miracle is that the subtraction is in the d's and does not impair the high relative accuracy property. However **qd** does not guarantee high relative accuracy so long as q's are dominated by neighbouring e's.

Since no intermediate quantities exceed σ_1, it is assumed that the initial data are scaled so that σ_1 (or σ_1^2 for **dqds**) is close to the overflow threshold. Underflow, though possible, is then a rare event.

Finally we remind the reader that the symbol $=$ carries its normal mathematical meaning.

7.2 High Relative Accuracy in the Presence of Shifts

We refer the reader to Section 5.3 where almost positive bidiagonals are introduced. Rutishauser merges the q's and e's into a single array Z;

$$Z := \{q_1, e_1, q_2, e_2, \ldots, e_{n-1}, q_n\}$$

and this is a convenient notation for the analysis which follows.

Before stating our claim we need more notation because the difficulty in the analysis is one of interpretation. Given Z the **dqds** algorithm in finite precision arithmetic produces representable output $\hat{Z}$. We introduce two ideal arrays $\vec{Z}$ and $\breve{Z}$ such that $\breve{Z}$ is the output of **dqds** with shift τ acting on $\vec{Z}$ *in exact arithmetic.* Moreover $\vec{Z}$ is a small relative perturbation of Z and $\hat{Z}$ is a small relative perturbation of $\breve{Z}$. See Figure 1.

Our model of arithmetic is that the floating point result of a basic arithmetic operation $\circ$ satisfies

$$fl(x \circ y) = (x \circ y)(1+\eta) = (x \circ y)/(1+\delta) \tag{17}$$

where η and δ depend on x, y, and $\circ$, and the arithmetic unit but satisfy

$$|\eta| < \epsilon \ , \quad |\delta| < \epsilon$$

for a given ϵ that depends only on the arithmetic unit. We shall choose freely the form (η or δ) that suits the analysis.

A fairly simple result is possible because the only truly sequential part of **dqds** is the sequence $\{d_i\}_1^n$. Note that, in exact arithmetic

$$d_{k+1} = \frac{d_k q_{k+1}}{d_k + e_k} - \tau^2$$

The trick is to write down the relations governing the computed quantities and then to *discern* among them an exact **dqds** transform whose input is close to Z and whose output is close to $\hat{Z}$.

Theorem 7.1. *In the absence of underflow or overflow, the Z diagram given above commutes and $\vec{q}_k$ ($\vec{e}_k$) differs from q_k (e_k) by 3 (1) ulps, $\hat{q}_k$ ($\hat{e}_k$) differrs from $\breve{q}_k$ ($\breve{e}_k$) by 2 (2) ulps.*

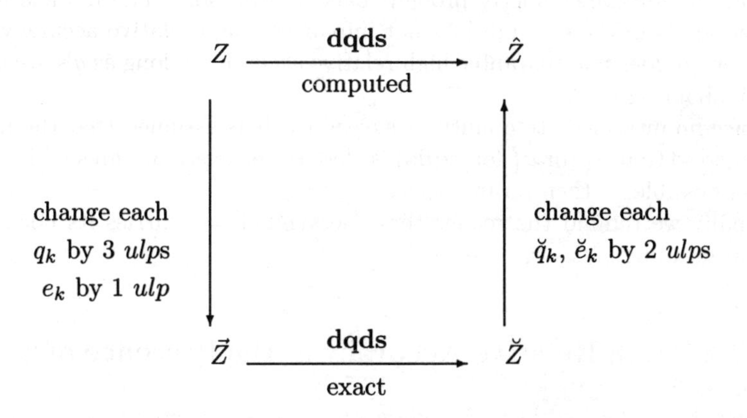

Figure 1: Effects of roundoff

Proof. We write down the exact relations satisfied by the computed quantities $\hat{Z}$.

$$\hat{q}_k = (d_k + e_k)/(1 + \epsilon_+) \tag{18}$$

$$t_k = q_{k+1}(1 + \epsilon_/)/\hat{q}_k = \frac{q_{k+1}(1+\epsilon_/)(1+\epsilon_+)}{d_k + e_k} \tag{19}$$

$$\hat{e}_k = e_k t_k (1 + \epsilon_*) = \frac{e_k q_{k+1}(1+\epsilon_/)(1+\epsilon_+)(1+\epsilon_*)}{d_k + e_k} \tag{20}$$

$$d_{k+1} = \frac{\{d_k t_k (1 + \epsilon_\star) - \tau^2\}}{1 + \epsilon_{k+1}}$$

Note the difference between $*$ and $\star$. Of course all the ϵ's obey (17) and depend on k but we have chosen to single out the one that accounts for the subtraction because it is the only one where the k dependence must be made explicit. In more detail the last relation is

$$\begin{aligned}(1+\epsilon_{k+1})d_{k+1} &= \frac{d_k q_{k+1}}{d_k + e_k}(1+\epsilon_/)(1+\epsilon_+)(1+\epsilon_\star) - \tau^2 \\ &= \frac{(1+\epsilon_k)d_k q_{k+1}(1+\epsilon_/)(1+\epsilon_+)(1+\epsilon_\star)}{(1+\epsilon_k)d_k + (1+\epsilon_k)e_k} - \tau^2 \end{aligned} \tag{21}$$

This tells us how to define $\vec{Z}$. Note that ϵ_k arose in the previous step. Moreover

$$(1 + \epsilon_1)d_1 = q_1 - \tau^2 \tag{22}$$

Our choice of $\vec{Z}$, in general, is not a machine representable array.

For $k \geq 1$,

$$\vec{d}_k := (1+\epsilon_k)d_k$$
$$\vec{e}_k := (1+\epsilon_k)e_k \tag{23}$$
$$\vec{q}_{k+1} := q_{k+1}(1+\epsilon_/)(1+\epsilon_+)(1+\epsilon_\star)\ ,\ \ (\vec{q}_1 = q_1) \tag{24}$$

and by (21),

$$\vec{d}_{k+1} = \frac{\vec{d}_k \vec{q}_{k+1}}{\vec{d}_k + \vec{e}_k} - \tau^2 \tag{25}$$

Then, for exact **dqds**, we must define

$$\breve{q}_k := \vec{d}_k + \vec{e}_k = (1+\epsilon_k)(d_k + e_k)$$
$$\breve{e}_k := \frac{\vec{e}_k \vec{q}_{k+1}}{\vec{d}_k + \vec{e}_k}.$$

Finally $\hat{q}_k$ and $\hat{e}_k$ must be recast in terms of $\breve{Z}$;

$$\hat{q}_k = \breve{q}_k/(1+\epsilon_k)(1+\epsilon_+)\ ,\ \ \text{from (18)} \tag{26}$$

$$\hat{e}_k = \frac{e_k q_{k+1}}{d_k + e_k}(1+\epsilon_/)(1+\epsilon_+)(1+\epsilon_*)\ ,\ \ \text{from (20)}$$
$$= \frac{\vec{e}_k \vec{q}_{k+1}}{(\vec{d}_k + \vec{e}_k)} \frac{(1+\epsilon_*)}{(1+\epsilon_\star)}. \tag{27}$$

It is (23) that yields $\vec{e}_k/(\vec{d}_k + \vec{e}_k) = e_k/(d_k + e_k)$. Equations (23) and (24) give the change from $\breve{Z}$ to $\hat{Z}$, and equation (25) fixes the exact **dqds** transform. □

Recall that, in exact arithmetic, algorithm **dqds** diminishes all eigenvalues (of LR) by the shift. For finite precision execution we have the following.

Corollary 7.1. *Algorithm* **dqds** *preserves high relative stability. When Z and $\hat{Z}$ are positive then the associated bidiagonal matrices B and B ($a_i = \sqrt{q_i}$, $b_i = \sqrt{e_i}$, etc.), together with the associated ideal bidiagonals $\vec{B}$ and $\breve{B}$, satisfy*

$$\sigma_i[\vec{B}] = \sigma_i[B] \exp\{2(n-1)\epsilon_1\},$$
$$\sigma_i^2[\breve{B}] = \sigma_i^2[\vec{B}] - \tau^2,$$
$$\sigma_i[\hat{B}] = \sigma_i[\breve{B}] \exp\{(2n-1)\epsilon_2\},$$

for $i = 1, 2, \ldots, n$, and $\epsilon_1 \leq \epsilon$, $\epsilon_2 \leq \epsilon$.

Proof. For $i = 1, \ldots, n-1$

$$\vec{a}_{i+1} = \sqrt{\vec{q}_{i+1}} = a_{i+1}\sqrt{(1+\epsilon_/)(1+\epsilon_+)(1+\epsilon_\star)}$$
$$\vec{b}_i = \sqrt{\vec{e}_i} = b_i\sqrt{(1+\epsilon_i)}.$$

By Theorem 2 in DK, the relative change in any singular value in going from B to $\vec{B}$ is the product of all the relative changes, namely

$$\prod_{i=1}^{n-1}[(1+\epsilon_/)(1+\epsilon_+)(1+\epsilon_\star)(1+\epsilon_i)]^{\frac{1}{2}} \leq \exp 2(n-1)\epsilon.$$

Similarly

$$\hat{a}_i = \sqrt{\hat{q}_i} = \breve{a}_i/\sqrt{(1+\epsilon_i)(1+\epsilon_+)} \ , \ i < n$$
$$\hat{b}_i = \sqrt{\hat{e}_i} = \breve{b}_i\sqrt{(1+\epsilon_*)(1+\epsilon_\star)} \ , \ i < n$$
$$\hat{a}_n = \sqrt{d_n} = \breve{a}_n/\sqrt{(1+\epsilon_n)}.$$

The relative change in any singular value in the transformation from $\breve{B}$ to $\hat{B}$ is bounded by

$$\sqrt{1+\epsilon_n}\prod_{i=1}^{n-1}[(1+\epsilon_*)/(1+\epsilon_\star)(1+\epsilon_i)(1+\epsilon_+)]^{\frac{1}{2}} \leq \exp{(4n-3)\epsilon/2}.$$

Since the passage from $\vec{B}$ to $\breve{B}$ is exact the singular values diminish by τ^2. □
Remark: It can be shown by similar means that one **dqd** transformation cannot alter any singular value by more than $3(n-1)$ *ulps*.

Theorem 4 is much stronger than the familiar error analysis based on norms because:

1. The perturbed matrices considered here inherit the bidiagonal structure

2. The bounds are very much smaller than those from DK (see Section 10) or the Golub-Reinsch algorithm (see Chapter 8 of [14]).

For multiple sweeps of **dqds**, our results can be stated more simply in terms of the positive sequence $\{\breve{Z}_l\}$ where l denotes the sweep with $\breve{Z}_1 = Z_1 = Z$. See Figure 2 for the corresponding commutative diagram. Then by combining $\breve{Z}_l \rightarrow Z_{l+1} \rightarrow \breve{Z}_{l+1}$ one obtains that $\breve{Z}_{l+1}$ is the exact **dqds** transformation of a perturbed (in relative sense) $\breve{Z}_l$. Thus backward stability is present for $\{\breve{Z}_l\}$.

Similarly it can be shown that the sequence $\{\vec{Z}_l\}$ is forward stable (in relative sense) with $\vec{Z}_f = Z_f$ where Z_f is the final computed result. On exact application of **dqds** on $\vec{Z}_l$ we get $\breve{Z}_{l+1}$ instead of $\vec{Z}_{l+1}$ (see Figure 2) and the error between $\vec{Z}_{l+1}$ and $\breve{Z}_{l+1}$ is small.
Example 3: The following experiment shows vividly the difference between an algorithm that obtains high relative accuracy (**dqds**) and one that does not (LINPACK **dsvdc** based on the Golub-Reinsch algorithm) but which delivers excellent absolute accuracy. We took the graded matrix B_+,

$$a_{i-1} = \beta a_i \quad , \quad b_i = a_i$$

with $a_n = 1$, $n = 8$ and $\beta = 60$. We applied both algorithms to B_+ and its reversal B_-,

$$a_i \rightarrow a_{n+1-i} \quad , \quad b_i \rightarrow b_{n-i}.$$

We did not allow any flipping of the matrix within the **dqds** algorithm although such flipping improves convergence. See next section.

In Tables 5 and 6, the third column shows, abs_i,

$$abs_i := (\sigma_i[B_-] - \sigma_i[B_+])/\sigma_1[B_+]$$

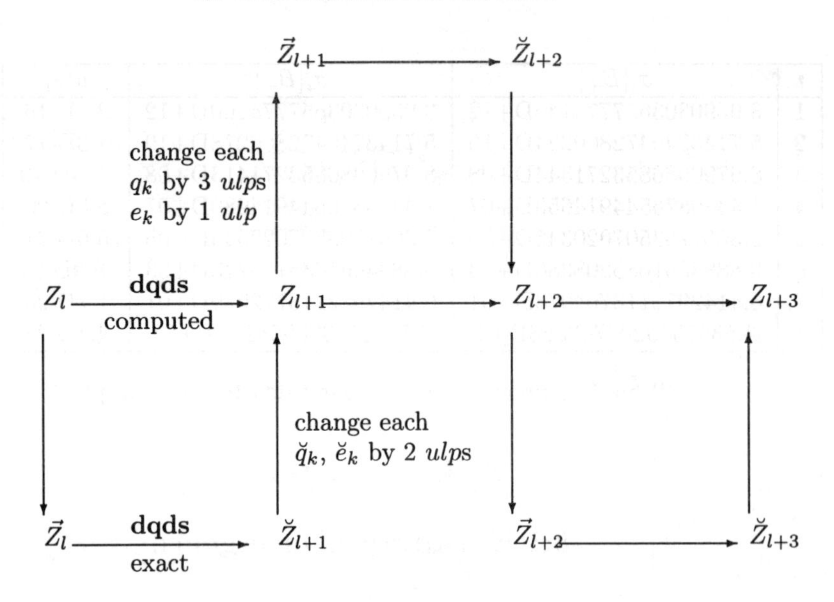

Figure 2: Effects of roundoff for multiple sweeps

the differences between outputs scaled by the 2-norm of the nicer matrix. Recall that *macheps* $\approx 2^{-53} \approx 1.1 \times 10^{-16}$. For **dsvdc** (see Table 5), it can be seen that the absolute error is even smaller than absolute stability guarantees.

In Tables 5 and 6, the fourth column shows, rel_i,

$$rel_i := (\sigma_i[B_-] - \sigma_i[B_+])/\sigma_i[B_+]$$

the relative differences in the outputs. For **dqds** the largest magnitude is less than two *macheps* (see Table 5) while for **dsvdc** (see Table 6), it is very much larger and shows that **dsvdc** does not give relative accuracy.

i	$\sigma_i[B_+]$	$\sigma_i[B_-]$	abs_i	rel_i
1	3.9590303657774160D+12	3.9590303657774155D+12	-1.2D-16	-1.2D-16
2	5.7143240472800255D+10	5.7143240472800247D+10	-1.9D-18	-1.3D-16
3	8.9790986853271568D+08	8.9790986853271568D+08	0.0D+00	0.0D+00
4	1.4489876544914651D+07	1.4489876544914651D+07	0.0D+00	0.0D+00
5	2.3661793507020348D+05	2.3661793507020348D+05	0.0D+00	0.0D+00
6	3.8884661685208386D+03	3.8884661685208386D+03	-1.1D-25	-1.2D-16
7	6.4142972113704085D+01	6.4142972113704109D+01	3.6D-27	2.2D-16
8	3.5351579203702068D-01	3.5351579203702068D-01	0.0D+00	0.0D+00

Table 5: Numerical results using **dqds** for Example 3

i	$\sigma_i[B_+]$	$\sigma_i[B_-]$	abs_i	rel_i
1	3.9590303657774146D+12	3.9590303657774160D+12	3.7D-16	3.7D-16
2	5.7143240472800224D+10	5.7143240472800278D+10	1.3D-17	9.3D-16
3	8.9790986853271544D+08	8.9790986853271413D+08	-3.3D-19	-1.5D-15
4	1.4489876544914653D+07	1.4489876544914989D+07	8.5D-20	2.3D-14
5	2.3661793507020345D+05	2.3661793507022545D+05	5.6D-21	9.3D-14
6	3.8884661685208386D+03	3.8884661685173243D+03	-8.9D-22	-9.0D-13
7	6.4142972113704073D+01	6.4142972113772929D+01	1.7D-23	1.1D-12
8	3.5351579203702068D-01	3.5351579205582154D-01	4.7D-24	5.3D-11

Table 6: Numerical results using **dsvdc** for Example 3

8 Convergence

8.1 Quadratic Convergence

Consider the last few steps in **dqds** with shift τ:

$$\hat{q}_{n-1} = d_{n-1} + e_{n-1}$$
$$\hat{e}_{n-1} = e_{n-1}q_n/\hat{q}_{n-1}$$
$$d_n = d_{n-1}q_n/\hat{q}_{n-1} - \tau^2$$
$$\hat{q}_n = d_n.$$

Hence

$$\begin{aligned} \hat{e}_{n-1}\hat{q}_n &= \frac{e_{n-1}q_n}{\hat{q}_{n-1}}\left[q_n(1-\frac{e_{n-1}}{\hat{q}_{n-1}})-\tau^2\right], \\ &= \frac{e_{n-1}q_n}{\hat{q}_{n-1}}\left[q_n-\tau^2-\frac{q_ne_{n-1}}{\hat{q}_{n-1}}\right] \end{aligned} \tag{28}$$

In exact arithmetic, as $\tau \to \sigma_n[B_n]$ we have $q_n \to 0$, $e_{n-1} \to 0$, $q_{n-1} \to \sigma_{n-1}^2[B_n] - \sigma_n^2[B_n] := gap > 0$ because the singular values are distinct if the initial B is of positive type. Thus convergence will be quadratic with respect to this *gap*.

Expression (28) shows that if

$$0 \le q_n - \tau^2 \le 2\frac{q_ne_{n-1}}{\hat{q}_{n-1}} \tag{29}$$

then

$$|q_n - \tau^2 - \frac{q_ne_{n-1}}{\hat{q}_{n-1}}| \le \frac{q_ne_{n-1}}{\hat{q}_{n-1}}$$

and so, by (28),

$$\frac{|\hat{e}_{n-1}\hat{q}_n|}{(e_{n-1}q_n)^2} \le \frac{1}{\hat{q}_{n-1}^2} \to \frac{1}{gap^2} > 0,$$

as $\tau \to \sigma_n[B_n]$. Thus (29) shows a (theoretical) interval for those τ that deliver quadratic convergence. Next we seek a usable expression that will ultimately lie in that interval.

The perfect shift is

$$\tau^2 = q_n(1 - \frac{e_{n-1}}{\hat{q}_{n-1}})$$

so that the natural strategy is to estimate $\hat{q}_{n-1}$ from

$$\begin{aligned}\hat{q}_{n-1} &= (1 - \frac{e_{n-2}}{\hat{q}_{n-2}})q_{n-1} - \tau^2 - e_{n-1} \\ &\approx (1 - \frac{e_{n-2}}{q_{n-2}})q_{n-1},\end{aligned}$$

when e_{n-2} and e_{n-1} are small enough. We may assume that $n > 2$ and so may use

$$\tau^2 = q_n * (1 - e_{n-1}/q_{n-1} * \max\{\frac{1}{2}, 1 - e_{n-2}/q_{n-2}\}) \tag{30}$$

provided that $0 < d_{n-1} = \min_{1 \le i \le n-1} d_i$ and $d_n < d_{n-1}$.

8.2 Cubic Convergence

The assertion in Section 8.1 that the shift $\tau^2 = q_n$ yields quadratic convergence for **qds** and **dqds** appears to contradict the result of Rutishauser [4] that this choice yields cubic convergence. See also Rutishauser and Schwarz [7] and Chapter 8 of Wilkinson [8]. Actually, there is no anomaly because the shift strategies are not quite the same. In our terminology what Rutishauser suggests is that the qd transform with $\tau^2 = q_n$ should not be formed explicitly. The only item wanted from it is $\hat{q}_n$ and it is assumed to be the only negative q_i. Then it is shown that $q_n + \hat{q}_n$ is a fourth order approximation to σ_n^2 *from below.* A qd transform of $\{q, e\}$ with shift $\tau^2 = q_n + \hat{q}_n$ will yield cubic convergence.

The point that is stressed by neither Rutishauser nor Wilkinson is that the computation of q_n costs $\mathcal{O}(n)$ operations, very close to 1 step of **qds**. From another perspective Rutishauser's analysis is a disguised derivation of the cubic convergence of the tridiagonal QR algorithm with the Rayleigh quotient shift.

For our algorithm **dqds** Rutishauser's (late failure) shift strategy described above is more appealing. Only the first phrase of our algorithm is needed to compute $\hat{q}_n$ and the cost is about $\frac{2}{3}$ of a **dqds** step. Moreover the positive form is preserved a little longer.

For the sake of completeness we indicate why $q_n + \hat{q}_n$ is a fourth order lower bound when q_n is second order in $q_n e_{n-1}$. The relevant tridiagonal matrix is BB^t and its leading principal $(n-1) \times (n-1)$ submatrix is called V. Recall that u_j is column j of I.

Fact 1: Provided $(V - \sigma_n^2 I)$ is invertible q_n may be written as

$$q_n = \sigma_n^2 + (q_n e_{n-1})^2 u_{n-1}^t (V - \sigma_n^2 I)^{-1} u_{n-1}$$

Fact 2: With shift $\tau^2 = q_n$,

$$\hat{q}_n = -(q_n e_{n-1})^2 u_{n-1}^t (V - q_n I)^{-1} u_{n-1}$$

Conclusion,

$$q_n + \hat{q}_n = \sigma_n^2 + (q_n e_{n-1})^2 u_{n-1}^t \left[(V - \sigma_n^2 I)^{-1} - (V - q_n I)^{-1}\right] u_{n-1}.$$

By Hilbert's second resolvent law,

$$q_n + \hat{q}_n = \sigma_n^2 + (q_n e_{n-1})^2 (\sigma_n^2 - q_n) u_{n-1}^t (V - \sigma_n^2 I)^{-1} (V - q_n I)^{-1} u_{n-1}.$$

Using Fact 1 again,

$$q_n + \hat{q}_n = \sigma_n^2 - (q_n e_{n-1})^4 \left\{ u_{n-1}^t (V - \sigma_n^2 I)^{-1} u_{n-1} u_{n-1}^t (V - q_n I)^{-1} (V - \sigma_n^2)^{-1} u_{n-1} \right\}.$$

The gap conditions ensure that the quantity in { } is $\mathcal{O}(1/gap)$.

In contrast to Rutishauser and Wilkinson, we have made no approximations in our derivation. The result is valid so long as the inverse matrices exist.

References

[1] H. Rutishauser. *Der quotienten-differenzen-algorithmus.* Z. Angew. Math. Phys., 5:233–251, 1954.

[2] H. Rutishauser. *Ein infinitesimales analogon zum quotienten-differenzen-algorithmus.* Arch. Math., 5:132–137, 1954.

[3] H. Rutishauser. *Solution of eigenvalue problems with the LR-transformation.* Nat. Bur. Standards Appl. Math. Series, 49:47–81, 1958.

[4] H. Rutishauser. *Über eine kubisch konvergente Variante der LR-Transformationa.* Z. Angew. Math. Mech., 11:49–54, 1960.

[5] H. Rutishauser. *Vorlesungen über numerische Mathematik.* Birkhäuser, Basel, 1976.

[6] H. Rutishauser. *Lectures on Numerical Mathematics.* Birkhäuser, Boston, 1990.

[7] H. Rutishauser and H. R. Schwarz. *The LR transformation method for symmetric matrices.* Numer. Math., 5:273–289, 1963.

[8] J. H. Wilkinson. *The Algebraic Eigenvalue Problem.* Clarendon Press, Oxford, 1965.

[9] B. Codenotti and M. Leoncini. *Parallel Complexity of Linear System Solution.* World Scientific, Singapore, 1991.

[10] J. Demmel and W. Kahan. *Accurate singular values of bidiagonal matrices.* SIAM J. Sci. Sta. Comput., 11:873–912, 1990.

[11] J. J. Dongarra, J. R. Bunch, C. B. Moler, and G. W. Stewart. *LINPACK User' Guide.* SIAM, Philadelphia, 1979.

[12] K. Vince Fernando and Beresford N. Parlett. *Orthogonal Cholesky algorithm.* Technical Report under preparation, Centre for Pure and Applied Mathematics, University of California at Berkeley, 1992.

[13] K. VINCE FERNANDO AND BERESFORD N. PARLETT. *qd algorithms for advanced architectures.* Technical Report under preparation, Centre for Pure and Applied Mathematics, University of California at Berkeley, 1992.

[14] G. H. GOLUB AND C. F. VAN LOAN. *Matrix Computations.* Johns Hopkins University Press, Baltimore, 1989.

[15] C. R. JOHNSON. *A Gersgorin-type lower bound for the smallest singular value.* Linear Algebra and Its Applications, 112:1–7, 1989.

Solution of shifted linear systems by quasi-minimal residual iterations

Roland W. Freund *

Abstract. High-order implicit methods for solving time-dependent partial differential equations and frequency response computations in control theory give rise to shifted systems of linear equations. Such systems have identical right-hand sides, and their coefficient matrices differ from each other only by scalar multiples of the identity matrix. This paper explores the use of two quasi-minimal residual iterations, the QMR and the TFQMR algorithm, for the solution of such shifted linear systems. It is shown that both algorithms can exploit the special structure, and that, for any family of shifted linear systems, the number of matrix-vector products and the number of inner products is the same as for a single linear system. Convergence results for the QMR and TFQMR algorithms are presented.

Key words. conjugate gradients, Krylov subspaces, quasi-minimal residual, transpose–free quasi–minimal residual.

AMS(MOS) subject classifications. 65F05, 65F10.

1 Introduction

Many iterative schemes for solving systems of linear equations fall into the category of *Krylov subspace methods*. Algorithms of this type use the coefficient matrix only in the form of matrix-vector products. Consequently, they are particularly well suited for the solution of large systems with sparse or structured dense coefficient matrices for which matrix-vector products can be computed cheaply. The most powerful Krylov subspace method is the classical *conjugate gradient algorithm* (CG) [16] for Hermitian positive definite systems. In recent years, there has been considerable interest in CG-like Krylov subspace methods for non-Hermitian linear systems, and a number of new algorithms were proposed. For a survey of these developments, we refer the reader to [9]. In particular, Krylov subspace methods with iterates characterized by a *quasi-minimal residual* property were developed. The first method of this type, the QMR algorithm, was proposed in [5]

*AT&T Bell Laboratories, Room 2C-420, 600 Mountain Road, Murray Hill, NJ 07974.

Numerical Linear Algebra

for the special case of complex symmetric matrices, and then extended to general non-Hermitian systems by Freund and Nachtigal [12]. The QMR algorithm is directly based on the nonsymmetric Lanczos process, and thus, like the latter, it requires matrix-vector products with both the coefficient matrix and its transpose. However, the products with the transpose only enter in the computation of certain auxiliary vectors that are needed to generate recurrence coefficients for the Lanczos process, while the vectors themselves are not used directly in the QMR algorithm. It is possible to devise quasi-minimal residual methods that do not require such auxiliary vectors and thus do not involve the transpose at all. A method of this type, the *transpose-free* QMR algorithm (TFQMR), was proposed in [7, 8]. We stress that QMR and TFQMR are not mathematically equivalent. In particular, both schemes generate different sets of iterates. The attractive feature of both algorithms, QMR and TFQMR, is that they generate *quasi-optimal* iterates with low work and storage requirements per iteration. In contrast, for non-Hermitian Krylov subspace methods that generate *optimal* iterates, such as the *generalized minimal residual algorithm* (GMRES) [21], work per iteration and storage requirements grow linearly with the iteration number. As a result, it is usually too expensive to run the full versions of these algorithms, and it is necessary to use restarted or truncated versions. For difficult problems, this often leads to very slow convergence.

In this paper, we explore the use of QMR and TFQMR for the solution of families of shifted linear systems of the form

$$(M - \sigma_j I)x^{(j)} = b, \quad j = 1, 2, \ldots, s. \tag{1.1}$$

Here $M \in \mathbb{C}^{N \times N}$ and $b \in \mathbb{C}^N$ are fixed, I is the $N \times N$ identity matrix, and $\sigma_j \in \mathbb{C}$, $j = 1, 2, \ldots, s$, are different shifts. We show that both algorithms, QMR and TFQMR, can exploit the special shift structure, and that, for any family of the type (1.1), the number of matrix-vector products and the number of inner products per iteration is the same as for a single linear system.

We remark that various authors studied the use of other Krylov subspace methods for the solution of shifted systems. Young [25] and Young and Vona [26] considered CG for the case of systems (1.1) with symmetric positive definite coefficient matrices $M - \sigma_j I$. In [4], we proposed and analyzed three different CG-type algorithms for the special class of shifted systems (1.1), where M is Hermitian and the σ_j's are general complex shifts. Finally, Datta and Saad [2] showed that the shift structure can be exploited when GMRES is used to solve general non-Hermitian systems (1.1).

Shifted linear systems (1.1) arise, for example, in the context of high-order implicit methods [1, 22] for solving time-dependent partial differential equations. Here, at each time step, one needs to solve a linear system of the form

$$\psi(M)x = \varphi(M)b, \tag{1.2}$$

where ψ and φ are polynomials with $\deg \varphi \leq \deg \psi =: s$. As pointed out in [24, 14], the solution of (1.2) can be obtained by solving a family of shifted linear systems (1.1). The idea is as follows. Let $\sigma_1, \sigma_2, \ldots, \sigma_s$ be the zeros of ψ, and assume that these zeros are all distinct. Then, expanding the rational function φ/ψ into partial fractions gives the decomposition

$$\frac{\varphi(\lambda)}{\psi(\lambda)} \equiv \mu_0 + \sum_{j=1}^{s} \frac{\mu_j}{\lambda - \sigma_j}, \tag{1.3}$$

where $\mu_0, \mu_1, \ldots, \mu_s$ are suitable scalars. Now assume that none of the poles $\sigma_1, \sigma_2, \ldots, \sigma_s$ in (1.3) coincides with an eigenvalue of M. Then, with (1.3), it follows that $x := (\psi(M))^{-1}\varphi(M)b$ is given by

$$x = \mu_0 b + \sum_{j=1}^{s} \mu_j x^{(j)}, \quad \text{where} \quad x^{(j)} := (M - \sigma_j I)^{-1} b.$$

Therefore, solving (1.2) is equivalent to solving a family of s shifted systems of the type (1.1). We remark that systems of the form (1.2) (with $\varphi(\lambda) \equiv 1$) also arise in connection with an algorithm proposed in [2] for constructing the Luenberger observer in control theory.

Another application from control theory that leads to shifted systems (1.1) is frequency response computations, see, e.g., [18]. The transfer matrix associated with continuous-time or discrete-time state space models is of the form

$$G(\sigma) \equiv C(\sigma I - M)^{-1}B + D, \quad \sigma \in \mathbb{C}. \tag{1.4}$$

Here $M \in \mathbb{C}^{N\times N}$, $C \in \mathbb{C}^{p\times N}$, $B \in \mathbb{C}^{N\times m}$, and $D \in \mathbb{C}^{p\times m}$. Frequency response computations for continuous-time models, respectively discrete-time models, require the evaluation of $G(\sigma)$ at points $\sigma_j = i\omega_j$, $\omega_j \in \mathbb{R}$, $j = 1, 2, \ldots, s$, on the imaginary axis, respectively at points $\sigma_j = e^{i\theta_j}$, $0 \le \theta_j < 2\pi$, $j = 1, 2, \ldots, s$, on the unit circle. Obviously, the columns of the corresponding matrices $(\sigma_j I - M)^{-1}B$ in (1.4) can be obtained by solving m families of shifted linear systems (1.1) with right hand-sides $b = b_k$, $k = 1, 2, \ldots, m$. Here the vectors b_k denote the columns of B. We remark that $m = 1$ for the important special case of single-input models.

The remainder of this paper is organized as follows. In Section 2, we briefly review the QMR and TFQMR algorithms. In Section 3, we discuss the use of QMR for shifted systems. In Section 4, we propose a version of TFQMR for shifted systems. In Section 5, we present error bounds for both algorithms. Finally, in Section 6, we make some concluding remarks.

Throughout the paper, all vectors and matrices are allowed to have real or complex entries. As usual, B^T and B^H denote the transpose and conjugate transpose of a matrix B, respectively. The vector norm $\|x\| := \sqrt{x^H x}$ is always the Euclidean norm, and $\|B\| := \max_{\|x\|=1} \|Bx\|$ is the associated matrix norm. For square matrices B, we denote by $\lambda(B)$ the set of eigenvalues of B. We use the notation

$$K_n(c, B) := \operatorname{span}\{c, Bc, \ldots, B^{n-1}c\}$$

for the nth Krylov subspace of $\mathbb{C}^N$ generated by $c \in \mathbb{C}^N$ and $B \in \mathbb{C}^{N\times N}$. We denote by

$$\mathcal{P}_n := \{\varphi(\lambda) \equiv \gamma_0 + \gamma_1\lambda + \cdots + \gamma_n\lambda^n \mid \gamma_0, \gamma_1, \ldots, \gamma_n \in \mathbb{C}\}$$

the set of all complex polynomials of degree at most n. We will make frequent use of the fact that

$$K_n(c, B) = \{\varphi(B)c \mid \varphi \in \mathcal{P}_{n-1}\}.$$

Finally, M is always an $N \times N$ matrix, $b \in \mathbb{C}^N$, and $\sigma_1, \sigma_2, \ldots, \sigma_s \in \mathbb{C}$. Moreover, we assume that

$$\sigma_j \notin \lambda(M) \quad \text{for all} \quad j = 1, 2, \ldots, s. \tag{1.5}$$

This guarantees that the coefficient matrices in (1.1) are all nonsingular.

2 The QMR and TFQMR algorithms

In this section, we briefly describe the QMR and TFQMR algorithms for solving single linear systems

$$Ax = b. \tag{2.1}$$

The QMR algorithm is directly based on a look-ahead variant of the classical nonsymmetric Lanczos process. First, we sketch this look-ahead Lanczos algorithm.

2.1 The look-ahead Lanczos algorithm

Let $A \in \mathbb{C}^{N \times N}$, and let $v_1, w_1 \in \mathbb{C}^N$, $v_1, w_1 \neq 0$, be arbitrary vectors. Starting with v_1 and w_1, the classical nonsymmetric Lanczos algorithm [17] generates two sequences of vectors $v_1, v_2, \ldots, v_n$ and $w_1, w_2, \ldots, w_n$ that span the Krylov subspaces $K_n(v_1, A)$ and $K_n(w_1, A^T)$, respectively. Furthermore, the two sequences are biorthogonal to each other, i.e., $w_j^T v_k = 0$ if $j \neq k$. Unfortunately, *breakdowns*—triggered by division by 0—and *near-breakdowns*—triggered by division by numbers close to 0—can occur in the classical Lanczos algorithm. Parlett, Taylor, and Liu [20] were the first to propose a modified procedure that uses so-called *look-ahead* techniques to avoid breakdowns and near-breakdowns. The QMR algorithm is based on a different implementation of the look-ahead Lanczos method that was recently developed by Freund, Gutknecht, and Nachtigal [10].

Starting with v_1 and w_1, this look-ahead Lanczos algorithm generates two sequences of vectors

$$v_1, v_2, \ldots, v_n \quad \text{and} \quad w_1, w_2, \ldots, w_n, \quad n = 1, 2, \ldots, \tag{2.2}$$

that, like the classical Lanczos process, satisfy

$$\begin{aligned} \operatorname{span}\{v_1, v_2, \ldots, v_n\} &= K_n(v_1, A), \\ \operatorname{span}\{w_1, w_2, \ldots, w_n\} &= K_n(w_1, A^T). \end{aligned} \tag{2.3}$$

However, the two sequences are now only block biorthogonal, i.e.,

$$(W^{(j)})^T V^{(k)} = 0 \quad \text{for all} \quad j \neq k, \quad j, k = 1, 2, \ldots, l(n). \tag{2.4}$$

The matrices $W^{(j)}$ and $V^{(k)}$ in (2.4) are defined by partitioning the Lanczos vectors (2.2) into blocks, according to the look-ahead steps taken:

$$[\, v_1 \quad v_2 \quad \cdots \quad v_n \,] = [\, V^{(1)} \quad V^{(2)} \quad \ldots \quad V^{(l(n))} \,],$$
$$[\, w_1 \quad w_2 \quad \cdots \quad w_n \,] = [\, W^{(1)} \quad W^{(2)} \quad \ldots \quad W^{(l(n))} \,].$$

More precisely, the matrices $V^{(k)}$ and $W^{(k)}$ are both of order $N \times h_k$, and their columns are just the Lanczos vectors corresponding to the kth look-ahead step. Here h_k denotes the length of the kth look-ahead step. Moreover, $l := l(n)$ denotes the number of look-ahead steps that were taken in the course of the first n steps of the Lanczos process. We denote by n_k the index of the first Lanczos vectors v_{n_k} and w_{n_k} in the kth blocks $V^{(k)}$ and $W^{(k)}$. In view of (2.4), the vectors v_{n_k} and w_{n_k} are biorthogonal to all previous Lanczos vectors $w_1, w_2, \ldots, w_{n_k-1}$ and $v_1, v_2, \ldots, v_{n_k-1}$, respectively. Therefore, the first vectors in each block are called *regular*. The remaining vectors in each block are called *inner*. We stress that the algorithm performs mostly standard Lanczos steps, i.e., look-ahead steps of size $h_k = 1$ with blocks $V^{(k)}$ and $W^{(k)}$ that consist of only single Lanczos vectors. True look-ahead steps, i.e., steps of size $h_k > 1$, are only used to avoid breakdowns and near-breakdowns. Typically, except for contrived examples, only few true look-ahead steps occur, and their size is usually small, mostly $h_k = 2$. We note that, if only steps of length $h_k = 1$ are performed, then the look-ahead Lanczos algorithm reduces to the classical algorithm.

We remark that the conditions (2.3) and (2.4) determine the Lanczos vectors only up to scaling. In the sequel, we always assume that the Lanczos vectors (2.2) are scaled to have unit length:

$$\|v_n\| = \|w_n\| = 1, \quad n = 1, 2, \ldots . \tag{2.5}$$

The crucial point of the look-ahead Lanczos process is that vectors satisfying (2.3) and (2.4) can be constructed by means of short recurrences, which involve only vectors from the last or the last two blocks. Next, we present a sketch of this look-ahead Lanczos algorithm.

Algorithm 2.1. (Sketch of the look-ahead Lanczos algorithm [10])

0) *Choose* $v_1, w_1 \in \mathbb{C}^N$ *with* $\|v_1\| = \|w_1\| = 1$.
   Set $V^{(1)} = v_1$, $W^{(1)} = w_1$, $D^{(1)} = (W^{(1)})^T V^{(1)}$, $n_1 = 1$, $l = 1$, $v_0 = w_0 = 0$, $V^{(0)} = v_0$, $W^{(0)} = w_0$, $D^{(0)} = \rho_1 = \xi_1 = 1$.

For $n = 1, 2, \ldots,$ *do*:

1) *Decide whether to construct* v_{n+1} *and* w_{n+1} *as regular or inner vectors and go to* 2) *or* 3), *respectively.*

2) (Regular step) *Compute*

$$\begin{aligned} \tilde{v}_{n+1} = A v_n &- V^{(l)} (D^{(l)})^{-1} (W^{(l)})^T A v_n \\ &- V^{(l-1)} (D^{(l-1)})^{-1} (W^{(l-1)})^T A v_n, \end{aligned} \tag{2.6}$$

$$\begin{aligned} \tilde{w}_{n+1} = A^T w_n &- W^{(l)} (D^{(l)})^{-T} (V^{(l)})^T A^T w_n \\ &- W^{(l-1)} (D^{(l-1)})^{-T} (V^{(l-1)})^T A^T w_n, \end{aligned}$$

set $n_{l+1} = n+1$, $l = l+1$, $V^{(l)} = W^{(l)} = \emptyset$, *and go to* 4).

3) (Inner step) *Compute*

$$\begin{aligned} \tilde{v}_{n+1} = Av_n &- \zeta_n v_n - (\eta_n/\rho_n)\, v_{n-1} \\ &- V^{(l-1)}(D^{(l-1)})^{-1}(W^{(l-1)})^T A v_n, \end{aligned} \tag{2.7}$$

$$\begin{aligned} \tilde{w}_{n+1} = A^T w_n &- \zeta_n w_n - (\eta_n/\xi_n)\, w_{n-1} \\ &- W^{(l-1)}(D^{(l-1)})^{-T}(V^{(l-1)})^T A^T w_n. \end{aligned} \tag{2.8}$$

4) *Compute* $\rho_{n+1} = \|\tilde{v}_{n+1}\|$ *and* $\xi_{n+1} = \|\tilde{w}_{n+1}\|$.
If $\rho_{n+1} = 0$ *or* $\xi_{n+1} = 0$: *set* $L = n$, *and stop. Otherwise, set*

$$\begin{aligned} v_{n+1} &= \tilde{v}_{n+1}/\rho_{n+1}, \quad w_{n+1} = \tilde{w}_{n+1}/\xi_{n+1}, \\ V^{(l)} &= [\, V^{(l)} \quad v_{n+1} \,], \quad W^{(l)} = [\, W^{(l)} \quad w_{n+1} \,], \quad D^{(l)} = (W^{(l)})^T V^{(l)}. \end{aligned}$$

We refer the reader to [10] for implementation details of Algorithm 2.1, and for a discussion of the criteria that are used for the decision in step 1).

We note that, in (2.7) and (2.8), $\zeta_n \in \mathbb{C}$ and $\eta_n \in \mathbb{C}$, $n = 0, 1, \ldots$, are arbitrary recurrence coefficients with $\eta_{n_k} = 0$, $k = 1, 2, \ldots$. The choice of these coefficients is not overly important, since Algorithm 2.1 usually performs only few inner steps anyway. We mostly used the algorithm with $\zeta_n := 1$ and, if $n \neq n_k$, $\eta_n := 1$.

Next, we list a few properties of Algorithm 2.1 that will be used later on; for proofs, we refer the reader to [10] and the references given there. Setting

$$V_n := [\, v_1 \quad v_2 \quad \cdots \quad v_n \,] \quad \text{and} \quad V_{n+1} := [\, V_n \quad v_{n+1} \,], \tag{2.9}$$

the recurrences (2.6), respectively (2.7), for the vectors $v_1, \ldots, v_n, v_{n+1}$, can be written compactly in matrix form as follows:

$$AV_n = V_{n+1} H_n. \tag{2.10}$$

Here H_n is an $(n+1) \times n$ upper Hessenberg matrix, which is also block tridiagonal with square blocks of order h_k on the diagonal. The subdiagonal elements of H_n are all nonzero, and hence

$$\operatorname{rank} H_n = n. \tag{2.11}$$

Each pair of Lanczos vectors v_n and w_n can be expressed in the form

$$v_n = \gamma_n \varphi_{n-1}(A) v_1 \quad \text{and} \quad w_n = \delta_n \varphi_{n-1}(A^T) w_1, \tag{2.12}$$

where $\varphi_{n-1} \in \mathcal{P}_{n-1}$ is of exact degree $n-1$ with leading coefficient 1, and $\gamma_n > 0$ and $\delta_n > 0$ are positive scalars. We refer to the φ_j's in (2.12) as the *look-ahead Lanczos polynomials*. Finally, we note that, in exact arithmetic, the stopping criterion in step 4) of Algorithm 2.1 will be satisfied after $L\,(\leq N)$ iterations, except in the very special situation of an incurable breakdown. The number L is called the *termination index*. An incurable breakdown occurs if Algorithm 2.1 starts to build infinite blocks $V^{(l)}$ and $W^{(l)}$ such that $(W^{(l)})^T V^{(l)}$ is the infinite zero matrix. Obviously, this is a very special situation, and we stress that incurable breakdowns do not pose a problem in practice. In the case of an incurable breakdown, we set $L := n_l - 1$, where n_l is the index of the last pair of regular vectors v_{n_l} and w_{n_l}.

We now turn to linear systems (2.1). For simplicity, from now on, we assume that the coefficient matrix A of (2.1) is nonsingular. We note that QMR and TFQMR generate well-defined iterates also for the case of singular A. A detailed discussion of QMR and TFQMR for singular systems can be found in [11].

2.2 The QMR method

Let $x_0 \in \mathbb{C}^N$ be any initial guess for the solution of (2.1), let $r_0 := b - Ax_0$ be the associated residual vector, and let $\tilde{r}_0 \in \mathbb{C}^N$ be an arbitrary nonzero vector. We now choose

$$v_1 := r_0/\|r_0\| \quad \text{and} \quad w_1 := \tilde{r}_0/\|\tilde{r}_0\| \tag{2.13}$$

as the starting vectors for Algorithm 2.1.

Using the matrices V_{n+1} and H_n generated by means of Algorithm 2.1, the QMR method [12] produces approximations x_n to the solution of (2.1) that are defined as follows:

$$x_n = x_0 + V_n z_n, \tag{2.14}$$

where $z_n \in \mathbb{C}^n$ is the solution of the least-squares problem

$$\|f_n - H_n z_n\| = \min_{z \in \mathbb{C}^n} \|f_n - H_n z\|. \tag{2.15}$$

The vector f_n is given by

$$f_n := \|r_0\| \cdot [1 \quad 0 \quad \cdots \quad 0]^T \in \mathbb{R}^{n+1}. \tag{2.16}$$

Note that, in view of (2.11), the solution z_n of (2.15) is unique. Furthermore, from (2.14), (2.9), (2.3), and (2.13), we have

$$x_n \in x_0 + K_n(r_0, A), \quad n = 1, 2, \ldots . \tag{2.17}$$

Using (2.14), (2.13), (2.16), and (2.10), one readily shows that the residual vector $r_n := b - Ax_n$ associated with x_n satisfies

$$r_n = V_{n+1}(f_n - H_n z_n). \tag{2.18}$$

Hence, in view of (2.15), the QMR iterates are characterized by a minimization of the second factor in the representation (2.18) of r_n; this is just the quasi-minimal residual property. Recall from (2.5) that the columns of V_{n+1} are scaled to be unit vectors. Therefore, the choice of z_n guarantees that all components in the representation (2.18) of r_n are treated equally in the quasi-minimization process. Other scaling strategies can also be used by simply defining z_n by a weighted least-squares problem, instead of (2.15); see [12] for details.

In the QMR algorithm, the least-squares problem (2.15) is solved by the standard approach based on a QR decomposition

$$H_n = Q_n^H \begin{bmatrix} R_n \\ 0 \end{bmatrix} \tag{2.19}$$

of H_n. Here Q_n is a unitary $(n+1)\times(n+1)$ matrix, and R_n is a nonsingular upper triangular $n \times n$ matrix. By means of (2.19), the least-squares problem (2.15) can be rewritten in the form

$$\min_{z \in \mathbb{C}^n} \|f_n - H_n z\| = \min_{z \in \mathbb{C}^n} \left\| Q_n f_n - \begin{bmatrix} R_n \\ 0 \end{bmatrix} z \right\|,$$

and thus z_n is given by

$$z_n = R_n^{-1} t_n, \quad \text{where} \quad t_n = \begin{bmatrix} \tau_1 \\ \vdots \\ \tau_n \end{bmatrix} \in \mathbb{C}^n, \quad \begin{bmatrix} t_n \\ \tilde{\tau}_{n+1} \end{bmatrix} := Q_n f_n. \tag{2.20}$$

Since H_n is upper Hessenberg, the unitary matrix Q_n can be chosen as a product of n Givens rotations. The QR decomposition (2.19) can then be updated easily from step to step. In particular, Q_n is obtained from Q_{n-1} by a simple multiplication with one Givens rotation that modifies the last two rows only. This implies that the vector t_n in (2.20) differs from the previous one, t_{n-1}, only by its additional last element τ_n. Together with (2.14) and (2.20), it follows that consecutive QMR iterates are connected by the update formula

$$x_n = x_{n-1} + \tau_n p_n, \quad \text{where} \quad p_n := V_n R_n^{-1} \begin{bmatrix} 0 \\ \vdots \\ 0 \\ 1 \end{bmatrix}. \tag{2.21}$$

The basic structure of the resulting QMR algorithm is then as follows.

Algorithm 2.2. (Sketch of the QMR algorithm [12])

0) *Choose* $x_0 \in \mathbb{C}^N$.
Set $r_0 = b - Ax_0$ *and* $v_1 = r_0/\|r_0\|$.
Choose $\tilde{r}_0 \in \mathbb{C}^N$ *with* $\tilde{r}_0 \neq 0$, *and set* $w_1 = \tilde{r}_0/\|\tilde{r}_0\|$.

For $n = 1, 2, \ldots,$ *do*:

1) *Perform the nth iteration of the look-ahead Lanczos Algorithm* 2.1.
This yields matrices V_n, V_{n+1}, *and* H_n *that satisfy* $AV_n = V_{n+1}H_n$.

2) *Update the* QR *factorization* (2.19) *of* H_n *and the vector* t_n *in* (2.20).

3) *Update the vector* p_n *in* (2.21).

4) *Set* $x_n = x_{n-1} + \tau_n p_n$.

5) *If* x_n *has converged*: *stop.*

The update of the vector p_n in step 3) can be implemented with only short recurrences. This is due to the block tridiagonal structure of H_n; see [12] for details. If no true look-ahead steps are taken, then H_n is even a scalar tridiagonal matrix, and the vectors p_n can be updated by simple three-term recurrences of the type

$$p_n = \varrho_n v_n - \epsilon_n p_{n-1} - \theta_n p_{n-2}, \tag{2.22}$$

where ϱ_n, ϵ_n, and θ_n are suitable scalars.

Finally, we remark that the procedure sketched in Algorithm 2.2 is not the only possible implementation of the QMR method. Recently, Freund and Nachtigal [13] devised an implementation that uses coupled two-term recurrences to generate the Lanczos vectors (2.2). The associated recurrence coefficients are just the entries of the block bidiagonal factors L_n and U_n of an LU decomposition $H_n = L_n U_n$ of the block tridiagonal matrix H_n. The QMR iterates are then obtained by solving a least-squares problem with L_n as coefficient matrix, instead of (2.15).

2.3 The TFQMR method

At each iteration step, the QMR method requires one matrix-vector multiplication with A, and one with A^T. However, only the products with A are used to advance the Krylov subspaces $K_n(r_0, A)$ from which, by (2.17), the QMR iterates are obtained. In [7, 8], we proposed a transpose-free quasi-minimal residual iteration, the TFQMR method, that exchanges each product with A^T for an additional product with A. As a result, with the same number of matrix-vector multiplications as QMR, the TFQMR method generates iterates from Krylov subspaces of twice the dimension. In this section, we sketch the TFQMR method.

Instead of one QMR iterate, the TFQMR method produces two iterates

$$x_m \in x_0 + K_m(r_0, A), \quad m = 2n-1,\, 2n, \tag{2.23}$$

in step n. Therefore, in the sequel, we will use $m = 1, 2, \ldots$, as the iteration index in our description of TFQMR. First, based on the look-ahead Lanczos polynomials given by (2.12), we define vectors

$$u_m := \begin{cases} \dfrac{\varphi_{n-1}(A)\varphi_n(A) r_0}{\|\varphi_{n-1}(A)\varphi_n(A) r_0\|}, & \text{if } m = 2n, \\[2ex] \dfrac{(\varphi_n(A))^2 r_0}{\|(\varphi_n(A))^2 r_0\|}, & \text{if } m = 2n+1, \end{cases} \tag{2.24}$$

and we set

$$U_m := \begin{bmatrix} u_1 & u_2 & \cdots & u_m \end{bmatrix}.$$

Since each polynomial φ_n is of exact degree n, we have

$$K_m(r_0, A) = \{ U_m z \mid z \in \mathbb{C}^m \}. \tag{2.25}$$

Furthermore, the vectors $u_1, u_2, \ldots, u_{m+1}$ satisfy recurrences of the form

$$A U_m = U_{m+1} F_m, \tag{2.26}$$

where F_m is an $(m+1) \times m$ upper Hessenberg matrix with nonzero subdiagonal elements. In particular, we have

$$\operatorname{rank} F_m = m. \tag{2.27}$$

Finally, note that, from (2.12), $\varphi_0(\lambda) \equiv 1$, and thus, by (2.24), $u_1 = r_0/\|r_0\|$.

Using (2.25), the iterates (2.23) can be parametrized in the form

$$x_m = x_0 + U_m z_m, \quad \text{where} \quad z_m \in \mathbb{C}^m. \tag{2.28}$$

Together with (2.26), it follows that the associated residual vector $r_m := b - Ax_m$ satisfies

$$r_m = U_{m+1}(f_m - F_m z_m), \tag{2.29}$$

where f_m is given by (2.16). In analogy to the definition (2.14)–(2.15) of the nth QMR iterate, the mth TFQMR iterate is given by (2.28) where z_m is chosen so that the Euclidean norm of the second factor in (2.29) is minimal, i.e., $z_m \in \mathbb{C}^m$ is the solution of the least-squares problem

$$\|f_m - F_m z_m\| = \min_{z \in \mathbb{C}^m} \|f_m - F_m z\|. \tag{2.30}$$

Note that, in view of (2.27), z_m is uniquely defined by (2.30).

Recall that the vectors u_m defined in (2.24) involve the polynomials (2.12) associated with the look-ahead Lanczos process. As a result, a robust implementation of TFQMR needs to have look-ahead incorporated in order to avoid breakdowns or near-breakdowns. It is an open question whether TFQMR with look-ahead can be implemented with only short vector recurrences and with only two matrix-vector multiplications by A per nth iteration. However, if no look-ahead is used, then an implementation with these features exists. Indeed, in [8], we derived an implementation of TFQMR without look-ahead that, instead of the vectors u_m, uses the search directions y_m from the *conjugate gradients squared algorithm* (CGS) [23] to update the TFQMR iterates. Like CGS, the resulting TFQMR algorithm without look-ahead can be implemented with only short vector updates and with only two matrix-vector multiplications per nth iteration. This implementation is as follows.

Algorithm 2.3. (TFQMR without look-ahead [8])

0) *Choose* $x_0 \in \mathbb{C}^N$.
   Set $\hat{u}_1 = y_1 = b - Ax_0$, $v_0 = Ay_1$, $d_0 = 0$.
   Set $\tau_0 = \|\hat{u}_1\|$, $\vartheta_0 = 0$, $\eta_0 = 0$.
   Choose $\tilde{r}_0 \in \mathbb{C}^N$ *such that* $\rho_0 = \tilde{r}_0^T \hat{u}_1 \neq 0$.

For $n = 1, 2, \ldots,$ *do*:

1) *If* $\tilde{r}_0^T v_{n-1} = 0$: *stop. Otherwise, set*

$$\alpha_{n-1} = \rho_{n-1}/(\tilde{r}_0^T v_{n-1}), \quad y_{2n} = y_{2n-1} - \alpha_{n-1} v_{n-1}. \tag{2.31}$$

2) *For* $m = 2n-1, 2n$ *do*:

   • *Set*

$$\begin{aligned} \hat{u}_{m+1} &= \hat{u}_m - \alpha_{n-1} A y_m, \\ \vartheta_m &= \|\hat{u}_{m+1}\|/\tau_{m-1}, \quad c_m = 1/\sqrt{1+\vartheta_m^2}, \\ \tau_m &= \tau_{m-1}\vartheta_m c_m, \quad \eta_m = c_m^2 \alpha_{n-1}, \\ d_m &= y_m + (\vartheta_{m-1}^2 \eta_{m-1}/\alpha_{n-1}) d_{m-1}, \\ x_m &= x_{m-1} + \eta_m d_m. \end{aligned} \tag{2.32}$$

- *If x_m has converged: stop.*

3) *Set* $\rho_n = \tilde{r}_0^T \hat{u}_{2n+1}$.
If $\rho_n = 0$: *stop. Otherwise, set*

$$\begin{aligned} \beta_n &= \rho_n/\rho_{n-1}, \quad y_{2n+1} = \hat{u}_{2n+1} + \beta_n y_{2n}, \\ v_n &= A y_{2n+1} + \beta_n (A y_{2n} + \beta_n v_{n-1}). \end{aligned} \tag{2.33}$$

We stress that the checks in step 1) and step 3) are necessary in order to avoid breakdowns in the form of division by 0. In general, it cannot be excluded that Algorithm 2.3 terminates in step 1) or 3), before x_m has converged. In an implementation of TFQMR with look-ahead, such premature termination of the algorithm would be avoided by using look-ahead techniques to skip over those steps in which division by 0 or by a number close to 0 would occur.

For later use, we list some properties of Algorithm 2.3; for proofs, we refer the reader to [8, 23]. The vectors $\hat{u}_m$ in (2.32) are rescaled versions of (2.24). More precisely,

$$\hat{u}_m := \begin{cases} \hat{\varphi}_{n-1}(A)\hat{\varphi}_n(A) r_0, & \text{if } m = 2n, \\ (\hat{\varphi}_n(A))^2 r_0, & \text{if } m = 2n+1, \end{cases} \tag{2.34}$$

where the $\hat{\varphi}_n$'s are rescaled versions of the Lanczos polynomials:

$$\hat{\varphi}_n(\lambda) \equiv \frac{\varphi_n(\lambda)}{\varphi_n(0)}. \tag{2.35}$$

We remark that the check in step 1) of Algorithm 2.3 guarantees that $\varphi_n(0) \neq 0$ in (2.35). The polynomials (2.35) can be generated by the following coupled two-term recurrences:

$$\begin{aligned} \psi_n(\lambda) &\equiv \hat{\varphi}_n(\lambda) + \beta_n \psi_{n-1}(\lambda), \\ \hat{\varphi}_{n+1}(\lambda) &\equiv \hat{\varphi}_n(\lambda) - \alpha_n \lambda \psi_n(\lambda), \end{aligned} \tag{2.36}$$

where $n = 0, 1, \dots$. For $n = 0$, we set $\hat{\varphi}_0(\lambda) \equiv 1$, $\beta_0 = 0$, and $\psi_{-1}(\lambda) \equiv 0$ in (2.36).

3 The QMR algorithm for shifted matrices

We now return to shifted linear systems of the form (1.1). From now on, we assume that M and the shifts $\sigma_1, \sigma_2, \dots, \sigma_s$ satisfy (1.5), so that the coefficient matrices $M - \sigma_j I$ in (1.1) are all nonsingular.

In the sequel, we always assume that $j \in \{1, 2, \dots, s\}$. We use the following convention. Quantities generated by Algorithm 2.1 or 2.2 applied to the shifted matrices $A = M - \sigma_j I$ are marked by superscripts $^{(j)}$, while quantities without superscripts denote scalars and vectors generated by the look-ahead Lanczos Algorithm 2.1 applied to the unshifted matrix $A = M$. For instance, $v_n^{(j)}$ and $w_n^{(j)}$ are the Lanczos vectors produced by Algorithm 2.1 applied to $M - \sigma_j I$, and v_n and w_n are the Lanczos vectors generated by Algorithm 2.1 applied to M. Finally, we always assume that all algorithms are started with the same initial vectors, independent of j. Hence, for Algorithm 2.1, we set

$$v_1^{(j)} = v_1 \quad \text{and} \quad w_1^{(j)} = w_1, \quad j = 1, 2, \ldots, s. \tag{3.1}$$

For the QMR algorithm, we set

$$x_0^{(j)} = 0, \quad r_0^{(j)} = r_0 = b, \quad \text{and} \quad \tilde{r}_0^{(j)} = \tilde{r}_0, \quad j = 1, 2, \ldots, s. \tag{3.2}$$

First, we compare the quantities generated by the look-ahead Lanczos Algorithm 2.1, when applied to $A = M$ and $A = M - \sigma_j I$. Obviously, in view of (3.1), we have, for all $n = 1, 2, \ldots$,

$$\begin{aligned} K_n(v_1^{(j)}, M - \sigma_j I) &= K_n(v_1, M), \\ K_n(w_1^{(j)}, (M - \sigma_j I)^T) &= K_n(w_1, M^T), \end{aligned} \tag{3.3}$$

and hence the Krylov subspaces (2.3) underlying the Lanczos process do not depend on the shifts σ_j. Furthermore, it is easily verified that the look-ahead Lanczos Algorithm 2.1 even generates identical basis vectors for the Krylov subspaces (3.3), provided the recurrence coefficients ζ_n in (2.7) and (2.8) are shifted accordingly. More precisely, we have the following result.

Lemma 3.1. *Let v_n and w_n, respectively $v_n^{(j)}$ and $w_n^{(j)}$, be the Lanczos vectors generated by Algorithm 2.1 applied to $A = M$, respectively $A = M - \sigma_j I$, with recurrence coefficients ζ_n and η_n, respectively $\zeta_n^{(j)} = \zeta_n - \sigma_j$ and $\eta_n^{(j)} = \eta_n$. Then the termination indices L and $L^{(j)}$ are the same, independent of the shifts σ_j, and*

$$v_n^{(j)} = v_n \quad \text{and} \quad w_n^{(j)} = w_n, \quad n = 1, 2, \ldots, L. \tag{3.4}$$

Furthermore, for $n = 1, 2, \ldots, L$,

$$(M - \sigma_j I)V_n = V_{n+1} H_n^{(j)}, \quad \text{where} \quad H_n^{(j)} := H_n - \sigma_j \begin{bmatrix} I_n \\ 0 \end{bmatrix}. \tag{3.5}$$

In (3.5), V_n and V_{n+1} are the matrices defined in (2.9), H_n is the upper Hessenberg matrix given by (2.10) (with $A = M$), and I_n denotes the $n \times n$ identity matrix.

We remark that, by means of (2.12), the relations (3.4) can be rewritten in terms of the look-ahead Lanczos polynomials, and we obtain

$$\varphi_n^{(j)}(\lambda - \sigma_j) \equiv \varphi_n(\lambda), \quad n = 0, 1, \ldots . \tag{3.6}$$

We now use the QMR Algorithm 2.2, with starting vectors (3.2), to solve each of the s linear systems (1.1). In view of Lemma 3.1, all s runs can be based on only one run of the look-ahead Lanczos Algorithm 2.1 (applied to M). A sketch of the resulting QMR process for solving a family of s shifted linear systems (1.1) is then as follows.

Algorithm 3.1. (QMR algorithm for solving s shifted linear systems (1.1))

0) *For $j = 1, 2, \ldots, s$, set $x_0^{(j)} = 0$ and $r_0^{(j)} = b$.*
   Set $v_1 = b/\|b\|$.
   Choose $\tilde{r}_0 \in \mathbb{C}^N$ with $\tilde{r}_0 \neq 0$, and set $w_1 = \tilde{r}_0/\|\tilde{r}_0\|$.

For $n = 1, 2, \ldots,$ *do*:

1) *Perform the nth iteration of the look-ahead Lanczos Algorithm* 2.1 *applied to* M. *This yields matrices* V_n, V_{n+1}, *and* H_n *that satisfy* $MV_n = V_{n+1}H_n$.

2) *For all* $j = 1, 2, \ldots, s$ *for which* $x_n^{(j)}$ *has not converged yet*:

   - *Update the* QR *factorization*
   $$H_n^{(j)} = (Q_n^{(j)})^H \begin{bmatrix} R_n^{(j)} \\ 0 \end{bmatrix}$$
   of the upper Hessenberg matrix
   $$H_n^{(j)} := H_n - \sigma_j \begin{bmatrix} I_n \\ 0 \end{bmatrix}.$$
   - *Update the vector*
   $$\begin{bmatrix} \tau_1^{(j)} \\ \vdots \\ \tau_n^{(j)} \\ \tilde{\tau}_{n+1}^{(j)} \end{bmatrix} = Q_n^{(j)} \begin{bmatrix} \|b\| \\ 0 \\ \vdots \\ 0 \end{bmatrix}.$$
   - *Update the vector*
   $$p_n^{(j)} = V_n (R_n^{(j)})^{-1} \begin{bmatrix} 0 \\ \vdots \\ 0 \\ 1 \end{bmatrix}.$$
   - *Compute*
   $$x_n^{(j)} = x_{n-1}^{(j)} + \tau_n^{(j)} p_n^{(j)}.$$

3) *If all* $x_n^{(j)}$ *have converged*: *stop.*

In Algorithm 3.1, matrix-vector products with M and M^T and inner products are computed only in step 1), while the vector updates in step 2) involve only SAXPYs. Consequently, for Algorithm 3.1 for solving s shifted linear systems, the number of matrix-vector products and the number of inner products per iteration is the same as for the QMR Algorithm 2.2 for solving a single linear system. Furthermore, we remark that, in view of (2.22), the vectors $p_n^{(j)}$ can be updated by simple three-term recurrences of the type
$$p_n^{(j)} = \varrho_n^{(j)} v_n - \epsilon_n^{(j)} p_{n-1}^{(j)} - \theta_n^{(j)} p_{n-2}^{(j)},$$
provided that no true look-ahead steps are taken.

Finally, we note that, for the applications described in Section 1, the matrix M and the right-hand side b are typically real, and only the shifts σ_j in (1.1) are in general complex. Obviously, in this case, the Lanczos vectors generated within Algorithm 2.1 are all real, as long as one chooses $w_1 \in \mathbb{R}^N$ and, in (2.7) and (2.8), real recurrence coefficients ζ_n and η_n. Therefore, even in the case of complex shifts, all quantities step 1) of Algorithm 3.1 are real. In particular, all matrix-vector products and all inner products involve only real vectors.

4 The TFQMR algorithm for shifted matrices

In this section, we derive a version of the TFQMR method for solving shifted linear systems. For simplicity, we only consider TFQMR without look-ahead, as implemented in Algorithm 2.3.

In the following, we always assume that $j \in \{1, 2, \ldots, s\}$. We continue to use the convention introduced in Section 3, and we use superscripts $^{(j)}$ to mark quantities associated with the TFQMR Algorithm 2.3 applied to the jth linear system $(M - \sigma_j I)x^{(j)} = b$ in the family (1.1). Vectors and scalars without superscripts denote quantities generated by the TFQMR Algorithm 2.3 applied to the unshifted matrix $A = M$, and φ_n, $\hat{\varphi}_n$, and ψ_n are the corresponding polynomials in (2.35) and (2.36). Finally, we assume that TFQMR is started with the initial vectors (3.2).

First, we note that, in view of (3.6), the normalized Lanczos polynomials (2.35), $\hat{\varphi}_n$ and $\hat{\varphi}_n^{(j)}$, satisfy

$$\hat{\varphi}_n(\lambda) \equiv \frac{\varphi_n(\sigma_j)}{\varphi_n(0)} \cdot \frac{\varphi_n^{(j)}(\lambda - \sigma_j)}{\varphi_n^{(j)}(0)} \equiv \gamma_n^{(j)} \hat{\varphi}_n^{(j)}(\lambda - \sigma_j), \tag{4.1}$$

$$\text{where} \quad \gamma_n^{(j)} := \hat{\varphi}_n(\sigma_j) = \frac{\varphi_n(\sigma_j)}{\varphi_n(0)}. \tag{4.2}$$

In addition to (4.2), we define

$$\delta_n^{(j)} := \psi_n(\sigma_j). \tag{4.3}$$

By (2.36) (with $\lambda = \sigma_j$), the quantities (4.2) and (4.3) can be updated as follows:

$$\begin{aligned} \delta_n^{(j)} &= \gamma_n^{(j)} + \beta_n \delta_{n-1}^{(j)}, \\ \gamma_{n+1}^{(j)} &= \gamma_n^{(j)} - \alpha_n \sigma_j \delta_n^{(j)}, \end{aligned} \tag{4.4}$$

where, for $n = 0$, we set $\gamma_0^{(j)} = 1$, $\beta_0 = 0$, and $\delta_{-1}^{(j)} = 0$. We will also need the scalars

$$\zeta_n^{(j)} := \hat{\varphi}_n^{(j)}(-\sigma_j) \quad \text{and} \quad \xi_n^{(j)} := \psi_n^{(j)}(-\sigma_j). \tag{4.5}$$

Setting $\lambda = 0$ in (4.1) and since, by (2.35), $\hat{\varphi}_n(0) = 1$, we have

$$\zeta_n^{(j)} = \frac{\hat{\varphi}_n(0)}{\hat{\varphi}_n(\sigma_j)} = \frac{1}{\gamma_n^{(j)}}. \tag{4.6}$$

Note that, in view of (2.36), the quantities (4.5) satisfy the relations

$$\xi_n^{(j)} = \zeta_n^{(j)} + \beta_n^{(j)} \xi_{n-1}^{(j)}, \tag{4.7}$$

$$\zeta_{n+1}^{(j)} = \zeta_n^{(j)} + \alpha_n^{(j)} \sigma_j \xi_n^{(j)}, \tag{4.8}$$

where, for $n = 0$, we set $\xi_{-1}^{(j)} = 0$ and $\beta_0^{(j)} = 0$ in (4.7).

After these preliminaries, we now begin our derivation of the TFQMR algorithm for shifted linear system. We will show that the TFQMR iterates $x_m^{(j)}$ for all s shifted systems (1.1) can be computed by means of simple vector updates from the vectors y_m, $\hat{u}_m$, and v_n generated by the TFQMR Algorithm 2.3 applied to the unshifted matrix $A = M$. Consequently, the resulting TFQMR method for s shifted systems requires per nth iteration only two inner products of vectors in $\mathbb{C}^N$, one norm computation $\|u\|$ of a vector $u \in \mathbb{C}^N$, and two matrix-vector multiplications with M; this is the same as for the TFQMR Algorithm 2.3 applied to a single system.

First, note that, from (4.1) and (4.6), we have

$$\hat{\varphi}_n^{(j)}(M - \sigma_j I) = \zeta_n^{(j)} \hat{\varphi}_n(M) \quad \text{for all} \quad n.$$

Together with (2.34) and (3.2), it follows that

$$\begin{aligned} \hat{u}_{2n}^{(j)} &= \zeta_{n-1}^{(j)} \zeta_n^{(j)} \hat{u}_{2n}, \\ \hat{u}_{2n+1}^{(j)} &= (\zeta_n^{(j)})^2 \hat{u}_{2n+1}. \end{aligned} \tag{4.9}$$

From the second relation in (4.9), we get

$$\beta_n^{(j)} = \frac{\tilde{r}_0^T \hat{u}_{2n+1}^{(j)}}{\tilde{r}_0^T \hat{u}_{2n-1}^{(j)}} = \left(\frac{\zeta_n^{(j)}}{\zeta_{n-1}^{(j)}} \right)^2 \beta_n. \tag{4.10}$$

By rewriting (4.8), we obtain

$$\alpha_n^{(j)} = \frac{\zeta_{n+1}^{(j)} - \zeta_n^{(j)}}{\sigma_j \xi_n^{(j)}}. \tag{4.11}$$

Note that the $\zeta_n^{(j)}$'s and $\xi_n^{(j)}$'s in (4.10) and (4.11) can be updated by means of the formulas (4.4), (4.7), and (4.6), which only involve the scalars α_n and β_n from the unshifted TFQMR algorithm. Consequently, the corresponding quantities $\alpha_n^{(j)}$ and $\beta_n^{(j)}$ for the shifted systems can be obtained by means of (4.11) and (4.10), respectively, without computing extra inner products.

Next, we consider the construction of the vectors $y_m^{(j)}$ that are needed for updating the search directions $d_m^{(j)}$ in step 2) of Algorithm 2.3. If we would generate all these vectors directly as in TFQMR for single systems, then this would require the computation of the matrix-vector products $(M - \sigma_j I) y_m^{(j)}$, which appear in the updates (2.32) and (2.33). We now show that this is actually not necessary. The key observation is that, from (2.32) (with $A = M - \sigma_j I$), we have

$$(M - \sigma_j I) y_m^{(j)} = \frac{1}{\alpha_{n-1}^{(j)}} \left(\hat{u}_m^{(j)} - \hat{u}_{m+1}^{(j)} \right), \quad m = 2n-1,\, 2n,\; n = 1, 2, \ldots . \tag{4.12}$$

Since, by (4.9), the $\hat{u}_m^{(j)}$'s are just rescaled versions of the $\hat{u}_m$'s from a TFQMR run for the unshifted matrix $A = M$, we can obtain the quantities $(M - \sigma_j I) y_m^{(j)}$ from (4.12), without computing additional matrix-vector products. Indeed, setting

$$\chi_n^{(j)} := \beta_n^{(j)} \frac{\alpha_n^{(j)}}{\alpha_{n-1}^{(j)}} \quad \text{and} \quad t_n^{(j)} := -\alpha_n^{(j)} v_n^{(j)},$$

and using (2.31), (2.33), (4.12), and (4.9), one readily verifies that the vectors $y_{2n-1}^{(j)}$ and $y_{2n}^{(j)}$ can be updated as follows:

$$
\begin{aligned}
y_{2n-1}^{(j)} &= (\zeta_{n-1}^{(j)})^2 \hat{u}_{2n-1} + \beta_{n-1}^{(j)} y_{2n-2}^{(j)}, \\
t_{n-1}^{(j)} &= \zeta_{n-1}^{(j)} \zeta_n^{(j)} \hat{u}_{2n} + (\chi_{n-1}^{(j)} - 1)(\zeta_{n-1}^{(j)})^2 \hat{u}_{2n-1} \\
&\quad - \chi_{n-1}^{(j)} (\zeta_{n-2}^{(j)} \zeta_{n-1}^{(j)} \hat{u}_{2n-2} - \beta_{n-1}^{(j)} t_{n-2}^{(j)}), \\
y_{2n}^{(j)} &= y_{2n-1}^{(j)} + t_{n-1}^{(j)}.
\end{aligned}
$$

Finally, the resulting TFQMR algorithm for solving s shifted linear systems (1.1) can be summarized as follows.

Algorithm 4.1. (TFQMR without look-ahead for shifted linear systems (1.1))

0) *Set* $\hat{u}_1 = y_1 = b$, $v_0 = My_1$, $\hat{u}_0 = 0$.
Choose $\tilde{r}_0 \in \mathbb{C}^N$ *such that* $\rho_0 = \tilde{r}_0^T \hat{u}_1 \neq 0$.
Set $\beta_0 = 0$.

1) *For* $j = 1, 2, \ldots, s$, *do*:

- *Set* $x_0^{(j)} = 0$.
- *Set* $\tau_0^{(j)} = \|\hat{u}_1\|$, $\vartheta_0^{(j)} = 0$, $\eta_0^{(j)} = 0$.
- *Set* $d_0^{(j)} = y_0^{(j)} = t_{-1}^{(j)} = 0$.
- *Set* $\gamma_0^{(j)} = \zeta_0^{(j)} = \zeta_{-1}^{(j)} = \alpha_{-1}^{(j)} = 1$, $\delta_{-1}^{(j)} = \xi_{-1}^{(j)} = 0$.

For $n = 1, 2, \ldots$ *do*:

2) *If* $\tilde{r}_0^T v_{n-1} = 0$: *stop. Otherwise, set*

$$\alpha_{n-1} = \rho_{n-1}/(\tilde{r}_0^T v_{n-1}), \quad y_{2n} = y_{2n-1} - \alpha_{n-1} v_{n-1}.$$

3) *For* $m = 2n-1,\, 2n$, *set*

$$\hat{u}_{m+1} = \hat{u}_m - \alpha_{n-1} M y_m.$$

4) *For all* $j = 1, 2, \ldots, s$ *for which* $x_m^{(j)}$ *has not converged yet*:

- *Set*

$$
\begin{aligned}
\delta_{n-1}^{(j)} &= \gamma_{n-1}^{(j)} + \beta_{n-1} \delta_{n-2}^{(j)}, \\
\gamma_n^{(j)} &= \gamma_{n-1}^{(j)} - \alpha_{n-1} \sigma_j \delta_{n-1}^{(j)}, \quad \zeta_n^{(j)} = 1/\gamma_n^{(j)}, \\
\beta_{n-1}^{(j)} &= (\zeta_{n-1}^{(j)}/\zeta_{n-2}^{(j)})^2 \beta_{n-1}, \\
\xi_{n-1}^{(j)} &= \zeta_{n-1}^{(j)} + \beta_{n-1}^{(j)} \xi_{n-2}^{(j)}, \\
\alpha_{n-1}^{(j)} &= (\zeta_n^{(j)} - \zeta_{n-1}^{(j)})/(\sigma_j \xi_{n-1}^{(j)}), \\
\chi_{n-1}^{(j)} &= \beta_{n-1}^{(j)} \alpha_{n-1}^{(j)} / \alpha_{n-2}^{(j)}.
\end{aligned}
$$

- *Set*

$$\begin{aligned} y_{2n-1}^{(j)} &= (\zeta_{n-1}^{(j)})^2 \hat{u}_{2n-1} + \beta_{n-1}^{(j)} y_{2n-2}^{(j)}, \\ t_{n-1}^{(j)} &= \zeta_{n-1}^{(j)} \zeta_n^{(j)} \hat{u}_{2n} + (\chi_{n-1}^{(j)} - 1)(\zeta_{n-1}^{(j)})^2 \hat{u}_{2n-1} \\ &\qquad -\chi_{n-1}^{(j)} (\zeta_{n-2}^{(j)} \zeta_{n-1}^{(j)} \hat{u}_{2n-2} - \beta_{n-1}^{(j)} t_{n-2}^{(j)}), \\ y_{2n}^{(j)} &= y_{2n-1}^{(j)} + t_{n-1}^{(j)}. \end{aligned}$$

- *For* $m = 2n-1,\ 2n$, *set*

$$\begin{aligned} \vartheta_m^{(j)} &= \frac{\|\hat{u}_{m+1}\|}{\tau_{m-1}^{(j)}} \zeta_n^{(j)} \cdot \begin{cases} \zeta_{n-1}^{(j)}, & \textit{if } m = 2n-1, \\ \zeta_n^{(j)}, & \textit{if } m = 2n, \end{cases} \\ c_m^{(j)} &= (1 + (\vartheta_m^{(j)})^2)^{-1/2}, \quad \tau_m^{(j)} = \tau_{m-1}^{(j)} \vartheta_m^{(j)} c_m^{(j)}, \\ d_m^{(j)} &= y_m^{(j)} + \left((\vartheta_{m-1}^{(j)})^2 \eta_{m-1}^{(j)} / \alpha_{n-1}^{(j)} \right) d_{m-1}^{(j)}, \\ \eta_m^{(j)} &= (c_m^{(j)})^2 \alpha_{n-1}^{(j)}, \quad x_m^{(j)} = x_{m-1}^{(j)} + \eta_m^{(j)} d_m^{(j)}. \end{aligned}$$

5) *If all* $x_m^{(j)}$ *have converged: stop.*

6) *Set* $\rho_n = \tilde{r}_0^T \hat{u}_{2n+1}$. *If* $\rho_n = 0$: *stop. Otherwise, set*

$$\begin{aligned} \beta_n &= \rho_n / \rho_{n-1}, \quad y_{2n+1} = \hat{u}_{2n+1} + \beta_n y_{2n}, \\ v_n &= M y_{2n+1} + \beta_n (M y_{2n} + \beta_n v_{n-1}). \end{aligned}$$

Like the QMR Algorithm 3.1, complex quantities can be avoided in the key steps of TFQMR Algorithm 4.1 when applied to shifted linear systems (1.1) with real M and b, but complex shifts σ_j. Indeed, if $\tilde{r}_0 \in \mathbb{R}^N$, then all scalars and vectors that are updated in steps 2), 3), and 6) remain real. In particular, all matrix-vector products and inner products involve only real vectors.

5 Convergence results

The quasi-minimal residual property of the QMR and TFQMR iterates is strong enough to deduce from it convergence results for both methods. We refer the reader to [12, 6] where convergence results are derived for QMR and TFQMR for solving single linear systems. In this section, we state these results for the QMR and TFQMR algorithms for shifted linear systems (1.1). For simplicity, these results are presented under some additional weak assumptions. We stress that these assumptions can be removed, and we remark that formulations of the convergence results for the general case can be found [6].

First, we state the convergence result for the QMR Algorithm 3.1. We will need some further notation. For any $(n+1) \times n$ matrix H_n, we denote by

$$\tilde{H}_n = [\, I_n \quad 0 \,] \, H_n \tag{5.1}$$

the matrix obtained from H_n by deleting its last row. Let L be the termination index of the look-ahead Lanczos Algorithm 2.1 (applied to $A = M$), as defined in Section 2.1. Let $\lambda_1, \lambda_2, \dots, \lambda_L$ be the zeros of the last look-ahead Lanczos polynomial φ_L, and let $\tilde{H}_L$ be the last $L \times L$ Lanczos matrix defined by (2.10) (with $A = M$) and (5.1). It is well known (see, e.g., [10] and the references given there) that

$$\{\lambda_1, \lambda_2, \dots, \lambda_L\} = \lambda(\tilde{H}_L) \subseteq \lambda(M). \tag{5.2}$$

For the remainder of this section, we assume that the zeros of φ_L are all distinct. In view of (5.2), this guarantees that the matrix $\tilde{H}_L$ is similar to the diagonal matrix

$$D_H := \operatorname{diag}(\lambda_1, \lambda_2, \dots, \lambda_L). \tag{5.3}$$

Recall from Lemma 3.1 that L is also the termination index for all shifted matrices $M - \sigma_j I$, $j = 1, 2, \dots, s$. Moreover, the last Lanczos matrices $\tilde{H}_L^{(j)}$ associated with the shifted systems are given by

$$\tilde{H}_L^{(j)} = H_L - \sigma_j I_L \quad \text{for all} \quad j = 1, 2, \dots, s. \tag{5.4}$$

From (5.4) and (5.2), we have

$$\{\lambda_1 - \sigma_j, \lambda_2 - \sigma_j, \dots, \lambda_L - \sigma_j\} = \lambda(\tilde{H}_L^{(j)}). \tag{5.5}$$

Moreover, by (5.4) and (5.3), there exists a nonsingular matrix $X \in \mathbb{C}^{L \times L}$ such that

$$X^{-1} \tilde{H}_L^{(j)} X = D_H - \sigma_j I_L \quad \text{for all} \quad j = 1, 2, \dots, s. \tag{5.6}$$

We stress that X does not depend on j.

By applying the QMR convergence result in [6, Theorem 5] to each of the single systems $(M - \sigma_j I) x^{(j)} = b$, and by using (5.5) and (5.6), we obtain the following theorem for the QMR Algorithm 3.1 for shifted systems.

Theorem 5.1. *Let L be the termination index of the look-ahead Lanczos Algorithm 2.1 (applied to $A = M$), and assume that the zeros $\lambda_1, \lambda_2, \dots, \lambda_L$ of φ_L are all distinct. Let D_H be the diagonal matrix defined in (5.3). Then, for $n = 1, 2, \dots, L-1$ and all $j = 1, 2, \dots, s$, the residual vectors $r_n^{(j)} := b - (M - \sigma_j I) x_n^{(j)}$ of the QMR Algorithm 3.1 for shifted linear systems (1.1) satisfy*

$$\frac{\|r_n^{(j)}\|}{\|b\|} \leq c_H \sqrt{n+1} \min_{\varphi \in \mathcal{P}_n:\, \varphi(0)=1} \max_{\lambda \in \lambda(M)} |\varphi(\lambda - \sigma_j)|,$$

$$\text{where} \quad c_H := \min_{X:\, X^{-1} \tilde{H}_L X = D_H} \|X\| \cdot \|X^{-1}\|.$$

Next, we turn to the TFQMR method for shifted linear systems. It can be shown that if look-ahead is used, then TFQMR terminates at iteration step $m = 2L$. Let $\tilde{F}_{2L}$ be the corresponding last TFQMR matrix as defined by (2.26) and (5.1). From the results in [6], it follows that

$$\{\lambda_1, \lambda_2, \dots, \lambda_L\} = \lambda(\tilde{F}_{2L}),$$

where each λ_k is an eigenvalue of algebraic multiplicity 2 and of geometric multiplicity 1. Hence $\tilde{F}_{2L}$ has the Jordan canonical form

$$J_F := \operatorname{diag}(J(\lambda_1), J(\lambda_2), \ldots, J(\lambda_L)), \quad \text{where} \quad J(\lambda_k) := \begin{bmatrix} \lambda_k & 1 \\ 0 & \lambda_k \end{bmatrix} \quad \text{for all} \quad k. \tag{5.7}$$

The convergence result for TFQMR can now be formulated as follows.

Theorem 5.2. *Let L be the termination index of the look-ahead Lanczos Algorithm* 2.1 *(applied to $A = M$), and assume that the zeros $\lambda_1, \lambda_2, \ldots, \lambda_L$ of φ_L are all distinct. Let J_F be the Jordan matrix defined in* (5.7). *Then, for $m = 1, 2, \ldots, 2L-1$ and all $j = 1, 2, \ldots, s$, the residual vectors $r_m^{(j)} := b - (M - \sigma_j I)x_m^{(j)}$ of the* TFQMR *method with look-ahead for shifted linear systems* (1.1) *satisfy*

$$\frac{\|r_m^{(j)}\|}{\|b\|} \leq c_F \sqrt{m+1} \min_{\varphi \in \mathcal{P}_m :\, \varphi(0)=1} \max_{\lambda \in \lambda(M)} \left(|\varphi(\lambda - \sigma_j)| + |\varphi'(\lambda - \sigma_j)|\right),$$

$$\textit{where} \quad c_F := \min_{X:\, X^{-1}\tilde{F}_{2L}X = J_F} \|X\| \cdot \|X^{-1}\|.$$

6 Concluding remarks

Families of linear systems with identical right-hand sides and shifted coefficient matrices arise in important applications. In this paper, we have derived special versions of two quasi-minimal residual iterations, the QMR and the TFQMR method, for solving such shifted linear systems. Both proposed algorithms take advantage of the special shift structure, and for any family of shifted systems, the number of matrix-vector products and the number of inner product is the same as for a single linear system.

Iterative methods for solving single linear systems are usually combined with a preconditioner to obtain faster convergence. Unfortunately, standard preconditioning techniques, such as incomplete factorization [19], destroy the special structure when they are applied to shifted linear systems. The only technique we are aware of that allows to preserve the shift structure is polynomial preconditioning. In [4], we studied polynomial preconditioning for the special class of linear systems (1.1) with Hermitian M and complex shifts. It turns out that, in this case, the optimal polynomial preconditioner is given by a suitable Chebyshev iteration [15] applied to the unshifted matrix M. Finally, we remark that Eiermann, Li, and Varga [3] developed a theory for polynomial preconditioning for asymptotically optimal Krylov subspace iterations for general non-Hermitian linear systems. In particular, their theory can also be used to analyze polynomial preconditioners for shifted linear systems.

Acknowledgements

I would like to thank Stratis Gallopoulos for pointing out some important references.

References

[1] J.C. CAVENDISH, W.E. CULHAM, AND R.S. VARGA, *A comparison of Crank-Nicolson and Chebyshev rational methods for numerically solving linear parabolic equations*, J. Comput. Phys. **10** (1972), 354–368.

[2] B.N. DATTA AND Y. SAAD, *Arnoldi methods for large Sylvester-like observer matrix equations, and an associated algorithm for partial spectrum assignment*, Linear Algebra Appl. **154–156** (1991), 225–244.

[3] M. EIERMANN, X. LI, AND R.S. VARGA, *On hybrid semiiterative methods*, SIAM J. Numer. Anal. **26** (1989), 152–168.

[4] R.W. FREUND, *On conjugate gradient type methods and polynomial preconditioners for a class of complex non-Hermitian matrices*, Numer. Math. **57** (1990), 285–312.

[5] R.W. FREUND, *Conjugate gradient-type methods for linear systems with complex symmetric coefficient matrices*, SIAM J. Sci. Statist. Comput. **13** (1992), 425–448.

[6] R.W. FREUND, *Quasi-kernel polynomials and convergence results for quasi-minimal residual iterations*, in: Numerical Methods of Approximation Theory, Vol. 9 (D. Braess and L.L. Schumaker, eds.), Birkäuser, Basel, 1992, pp. 77–95.

[7] R.W. FREUND, *Transpose-free quasi-minimal residual methods for non-Hermitian linear systems*, AT&T Numerical Analysis Manuscript 92–07, Bell Laboratories, Murray Hill, New Jersey, 1992.

[8] R.W. FREUND, *A transpose-free quasi-minimal residual algorithm for non-Hermitian linear systems*, SIAM J. Sci. Comput. **14** (1993), to appear.

[9] R.W. FREUND, G.H. GOLUB, AND N.M. NACHTIGAL, *Iterative solution of linear systems*, Acta Numerica **1** (1992), 57–100.

[10] R.W. FREUND, M.H. GUTKNECHT, AND N.M. NACHTIGAL, *An implementation of the look-ahead Lanczos algorithm for non-Hermitian matrices*, SIAM J. Sci. Comput. **14** (1993), to appear.

[11] R.W. FREUND AND M. HOCHBRUCK, *On the use of two* QMR *algorithms for solving singular systems and applications in Markov chain modeling*, Journal of Numerical Linear Algebra with Applications **2** (1993), to appear.

[12] R.W. FREUND AND N.M. NACHTIGAL, QMR: *a quasi-minimal residual method for non-Hermitian linear systems*, Numer. Math. **60** (1991), 315–339.

[13] R.W. FREUND AND N.M. NACHTIGAL, *An implementation of the* QMR *method based on coupled two-term recurrences*, in: Numerical Linear Algebra, L. Reichel, A. Ruttan and R.S. Varga, eds., de Gruyter, Berlin, 1993, pp 118-135.

[14] E. GALLOPOULOS AND Y. SAAD, *Efficient parallel solution of parabolic equations: implicit methods on the Cedar multicluster*, in: Proc. Fourth SIAM Conf. Parallel Processing for Scientific Computing (J. Dongarra, P. Messina, D.C. Sorensen, and R.G. Voigt, eds.), SIAM, Philadelphia, 1990, pp. 251–256.

[15] G.H. GOLUB AND R.S. VARGA, *Chebyshev semi-iterative methods, successive overrelaxation iterative methods, and second order Richardson iterative methods*, Numer. Math. **3** (1961), 147–168.

[16] M.R. HESTENES AND E.L. STIEFEL, *Methods of conjugate gradients for solving linear systems*, J. Res. Nat. Bur. Standards **49** (1952), 409–436.

[17] C. LANCZOS, *An iteration method for the solution of the eigenvalue problem of linear differential and integral operators*, J. Res. Nat. Bur. Standards **45** (1950), 255–282.

[18] A.J. LAUB, *Numerical linear algebra aspects of control design computations*, IEEE Trans. Automat. Contr. **AC-30** (1985), 97–108.

[19] J.A. MEIJERINK AND H.A. VAN DER VORST, *An iterative solution method for linear systems of which the coefficient matrix is a symmetric M-matrix*, Math. Comp. **31** (1977), 148–162.

[20] B.N. PARLETT, D.R. TAYLOR, AND Z.A. LIU, *A look-ahead Lanczos algorithm for unsymmetric matrices*, Math. Comp. **44** (1985), 105–124.

[21] Y. SAAD AND M.H. SCHULTZ, GMRES: *a generalized minimal residual algorithm for solving nonsymmetric linear systems*, SIAM J. Sci. Statist. Comput. **7** (1986), 856–869.

[22] W.L. SEWARD, G. FAIRWEATHER, AND R.L. JOHNSTON, *A survey of higher-order methods for the numerical integration of semidiscrete parabolic problems*, IMA J. Numer. Anal. **4** (1984), 375–425.

[23] P. SONNEVELD, CGS, *a fast Lanczos-type solver for nonsymmetric linear systems*, SIAM J. Sci. Statist. Comput. **10** (1989), 36–52.

[24] R.A. SWEET, *A parallel and vector variant of the cyclic reduction algorithm*, SIAM J. Sci. Statist. Comput. **9** (1988), 761–765.

[25] D.M. YOUNG, *The search for "high-level" parallelism for iterative sparse linear systems solvers*, in: Parallel Supercomputing: Methods, Algorithms and Applications (G.F. Carey, ed.), John Wiley, Chichester, 1989, pp. 89–105.

[26] D.M. YOUNG AND B.R. VONA, *On the use of rational iterative methods for solving large sparse linear systems*, Appl. Numer. Math. **10** (1992), 261–278.

Implementation details of the coupled QMR algorithm

Roland W. Freund *and Noël M. Nachtigal* †

Abstract. The original quasi-minimal residual method (QMR) relies on the three-term look-ahead Lanczos process to generate basis vectors for the underlying Krylov subspaces. However, empirical observations indicate that, in finite precision arithmetic, three-term vector recurrences are less robust than mathematically equivalent coupled two-term recurrences. Therefore, we recently proposed a new implementation of the QMR method based on a coupled two-term look-ahead Lanczos procedure. In this paper, we describe implementation details of this coupled QMR algorithm, and we present results of numerical experiments.

Key Words. conjugate gradient, Krylov subspaces, look–ahead Lanczos, quasi–minimal residual.

AMS(MOS) Subject Classifications. 65F05, 65F10.

1 Introduction

Recently, we proposed a new Krylov subspace iteration, the quasi-minimal residual algorithm (QMR) [5], for solving general nonsingular non-Hermitian systems of linear equations

$$Ax = b. \tag{1.1}$$

The QMR method has two main ingredients: the look-ahead Lanczos process, and a quasi-minimal residual condition. The look-ahead Lanczos algorithm is used to generate—with low work and storage requirements—basis vectors for the underlying Krylov subspaces. Furthermore, look-ahead techniques are used to avoid possible breakdowns in the classical Lanczos algorithm [7], except for so-called incurable breakdowns. Once the Lanczos basis is constructed, the quasi-minimal residual property is used to select the QMR iterates from the Krylov subspaces. As was shown in [5], the QMR iterates are always well defined, and the quasi-minimal

*AT&T Bell Laboratories, Room 2C-420, 600 Mountain Road, Murray Hill, NJ 07974.
†Research Institute for Advanced Computer Science, NASA Ames Research Center, Moffett Field, CA 94035.

residual condition leads to a smooth and nearly monotone convergence behavior. In addition, thanks to the quasi-minimal residual property, it is possible to prove convergence results for the QMR algorithm. The result is a method with several desirable numerical and theoretical properties.

In the original QMR algorithm, the look-ahead Lanczos method used generates the basis vectors for the Krylov subspaces by means of three-term recurrences. It has been observed that, in finite precision arithmetic, vector iterations based on three-term recursions are usually less robust than mathematically equivalent coupled two-term vector recurrences. Therefore, in [6], we proposed a new implementation of the QMR algorithm, based on a coupled two-term recurrence formulation of the Lanczos algorithm. Together with the derivation, we discussed in [6] the properties of the new implementation, and presented numerical results showing that the new method is more robust than the original QMR algorithm. In this paper, we discuss in more detail the implementation of the new algorithm; in particular, we focus on the implementation of the coupled two-term version of the look-ahead Lanczos algorithm. We also give several new numerical examples.

The outline of the paper is as follows. In Section 2, we recall the three-term look-ahead Lanczos process that was proposed in [3]. Then, in Section 3, we review the coupled two-term look-ahead Lanczos algorithm that was proposed in [6], and in Section 4 we give implementation details of the new algorithm. In Section 5, we briefly recall how the QMR approach can be combined with the coupled Lanczos algorithm to obtain a new implementation of the QMR method, and in Section 6, we report results of numerical experiments with this new QMR algorithm. Finally, in Section 7, we make some concluding remarks.

Throughout the paper, all vectors and matrices are allowed to have real or complex entries. As usual, $M^T = \begin{bmatrix} m_{kj} \end{bmatrix}$ denotes the transpose of the matrix $M = \begin{bmatrix} m_{jk} \end{bmatrix}$. We use $\sigma_{\min}(M)$ for the smallest singular value of M, while the vector norm $\|x\| := \sqrt{x^H x}$ is always the Euclidean norm. We denote by

$$K_n(c, B) := \operatorname{span}\{c, Bc, \ldots, B^{n-1}c\}$$

the nth Krylov subspace of $\mathbb{C}^N$ generated by $c \in \mathbb{C}^N$ and the $N \times N$ matrix B. Finally, it is always assumed that A is an $N \times N$ matrix, singular or nonsingular.

2 The three-term look-ahead Lanczos algorithm

The Lanczos process is a method that builds basis vectors for two Krylov subspaces, with low work and storage requirements. Given two starting vectors, v_1 and $w_1 \in \mathbb{C}^N$, the algorithm computes two sequences of vectors, $\{v_j\}_{j=1}^n$ and $\{w_j\}_{j=1}^n$, such that, for $n = 1, 2, \ldots,$

$$\begin{aligned} \operatorname{span}\{v_1, v_2, \ldots, v_n\} &= K_n(v_1, A), \\ \operatorname{span}\{w_1, w_2, \ldots, w_n\} &= K_n(w_1, A^T). \end{aligned} \tag{2.1}$$

In addition, the two sets of vectors are required to obey a biorthogonality relation. Ideally, one would like to impose the condition

$$w_n^T v_j = w_j^T v_n = 0 \quad \text{for all} \quad j < n. \tag{2.2}$$

This is done in the Lanczos process, as proposed by Lanczos in [7]. However, it turns out that it is not always possible or numerically stable to construct vectors satisfying (2.2), as *exact breakdowns* ($w_n^T v_n = 0$) or *near-breakdowns* ($w_n^T v_n$ is small in some sense) may arise. This poses a problem, since the construction of the pair v_{n+1} and w_{n+1} obeying (2.2) involves division by $w_n^T v_n$. As a remedy, one relaxes the biorthogonality condition, requiring instead that, in case of a breakdown, the relations (2.2) hold only for some range of j up to, but not equal to, n. This leads to so-called *look-ahead* Lanczos algorithms, which skip over the exact and near-breakdowns. The original QMR algorithm is based on a look-ahead Lanczos method proposed by Freund, Gutknecht, and Nachtigal [3], which we briefly review next.

Like the classical algorithm, the look-ahead Lanczos algorithm [3] generates vectors $\{v_j\}_{j=1}^n$ and $\{w_j\}_{j=1}^n$ with (2.1). In addition, they satisfy the biorthogonality relation

$$W_n^T V_n = D_n, \tag{2.3}$$

where D_n is a block diagonal matrix whose structure is discussed below, and

$$V_n := [\, v_1 \quad v_2 \quad \cdots \quad v_n \,] \quad \text{and} \quad W_n := [\, w_1 \quad w_2 \quad \cdots \quad w_n \,].$$

This means that some of the vectors $\{v_j\}_{j=1}^n$ and $\{w_j\}_{j=1}^n$ do in fact satisfy the full biorthogonality (2.2). These vectors are called *regular*, and they form a subsequence $\{v_{n_j}\}_{j=1}^l$ and $\{w_{n_j}\}_{j=1}^l$, where

$$1 =: n_1 < n_2 < \cdots < n_l \le n < n_{l+1}, \quad l := l(n). \tag{2.4}$$

All vectors that are not regular are called *inner*. The regular vectors are used to partition $\{v_j\}_{j=1}^n$ and $\{w_j\}_{j=1}^n$ into blocks. By convention, one defines blocks $V^{(j)}$, of size $N \times h_j$, containing the regular vector v_{n_j} and all inner vectors—if any—between v_{n_j} and $v_{n_{j+1}}$:

$$V^{(j)} = [\, v_{n_j} \quad v_{n_j+1} \quad \cdots \quad].$$

A look-ahead step is then defined in terms of building such a block. Hence the integer l in (2.4) is just the number of look-ahead steps taken during the first n steps of the Lanczos algorithm. Moreover, h_j is called the length of the jth look-ahead step. The structure of the sequence $\{w_j\}_{j=1}^n$ parallels that of the sequence $\{v_j\}_{j=1}^n$, so that V_n and W_n can be written as

$$V_n = [\, V^{(1)} \quad V^{(2)} \quad \cdots \quad V^{(l)} \,] \quad \text{and} \quad W_n = [\, W^{(1)} \quad W^{(2)} \quad \cdots \quad W^{(l)} \,],$$

and D_n in (2.3) is given by

$$D_n = \operatorname{diag}(D^{(1)}, D^{(2)}, \ldots, D^{(l)}), \quad D^{(j)} := (W^{(j)})^T V^{(j)}.$$

The choice of whether to build a regular or an inner vector is determined at each step, based on the particular look-ahead strategy used. It should also be pointed out that, even though the look-ahead Lanczos algorithm can handle most breakdowns, there remains a class of breakdowns, the so-called *incurable breakdowns*, which cannot be cured by look-ahead techniques. However, incurable breakdowns occur only in very particular circumstances, and they do not pose a problem in practice. Finally, since the scaling of the Lanczos vectors is not determined by (2.1) and (2.3), the look-ahead algorithm scales the vectors to have unit length:

$$\|v_n\| = \|w_n\| = 1, \quad n = 1, 2, \dots . \tag{2.5}$$

The main point of the look-ahead Lanczos process is that vectors satisfying (2.1) and (2.3) can be constructed by means of short block three-term recurrences. These recurrences can be written compactly as

$$AV_n = V_{n+1} H_n, \tag{2.6}$$

$$A^T W_n = W_{n+1} \Gamma_{n+1}^{-1} H_n \Gamma_n, \tag{2.7}$$

where H_n is an $(n+1) \times n$ block tridiagonal matrix,

$$\Gamma_n := \operatorname{diag}(\gamma_1, \gamma_2, \dots, \gamma_n), \quad \text{where} \quad \gamma_j := \begin{cases} 1, & \text{if} \quad j = 1, \\ \gamma_{j-1} \rho_j / \xi_j, & \text{if} \quad j > 1, \end{cases} \tag{2.8}$$

is a diagonal scaling matrix with positive diagonal entries, and ρ_j and ξ_j are scale factors used to ensure that v_j and w_j, respectively, obey the scaling (2.5). Since the recurrences used to build v_{n+1} and w_{n+1} are short, the look-ahead algorithm has low work and storage requirements, making it an attractive method for building bases for Krylov subspaces. However, it has been observed that, in finite precision arithmetic, three-term vector recurrences are less robust than mathematically equivalent coupled two-term recurrences. This was our motivation in proposing in [6] a different implementation of the look-ahead Lanczos algorithm, based on coupled two-term recurrences. Next, we review this algorithm.

3 The coupled two-term look-ahead Lanczos process

The coupled two-term Lanczos process is an alternate way of generating the Lanczos basis vectors.* The algorithm generates, in addition to the Lanczos vectors $\{v_j\}_{j=1}^n$ and $\{w_j\}_{j=1}^n$, a second set of basis vectors, $\{p_j\}_{j=1}^n$ and $\{q_j\}_{j=1}^n$, such that, for $n = 1, 2, \dots$,

$$\operatorname{span}\{p_1, p_2, \dots, p_n\} = K_n(v_1, A),$$
$$\operatorname{span}\{q_1, q_2, \dots, q_n\} = K_n(w_1, A^T).$$

For simplicity, we will sometimes refer to the Lanczos vectors $\{v_j\}_{j=1}^n$ and $\{w_j\}_{j=1}^n$ as the V–W sequence, and to the auxiliary vectors $\{p_j\}_{j=1}^n$ and $\{q_j\}_{j=1}^n$ as the P–Q

*The discussion that follows will not cover all the details and will not justify all the statements made. For full details, we refer the reader to [6].

sequence. The four sets of basis vectors are generated using coupled two-term recurrences of the form:

$$V_n = P_n U_n, \qquad AP_n = V_{n+1} L_n,$$

$$W_n = Q_n \Gamma_n^{-1} U_n \Gamma_n, \qquad A^T Q_n = W_{n+1} \Gamma_{n+1}^{-1} L_n \Gamma_n,$$

Here,

$$P_n := [\, p_1 \quad p_2 \quad \cdots \quad p_n \,] \quad \text{and} \quad Q_n := [\, q_1 \quad q_2 \quad \cdots \quad q_n \,],$$

while U_n is an upper triangular matrix and L_n is an upper Hessenberg matrix, given by

$$U_n := \begin{bmatrix} 1 & u_{12} & \cdots & u_{1n} \\ 0 & 1 & \ddots & \vdots \\ \vdots & \ddots & \ddots & u_{n-1,n} \\ 0 & \cdots & 0 & 1 \end{bmatrix} \quad \text{and} \quad L_n := \begin{bmatrix} l_{11} & l_{12} & \cdots & l_{1n} \\ \rho_2 & l_{22} & & \vdots \\ 0 & \rho_3 & \ddots & \vdots \\ \vdots & \ddots & \ddots & l_{nn} \\ 0 & \cdots & 0 & \rho_{n+1} \end{bmatrix},$$

and Γ_n is the diagonal matrix defined in (2.8). As was shown in [6], the matrices L_n and U_n define a factorization of the block tridiagonal Hessenberg matrix H_n generated by the three-term look-ahead Lanczos algorithm,

$$H_n = L_n U_n. \tag{3.1}$$

In addition, it is possible to reduce L_n and U_n to block bidiagonal matrices, by constructing the basis vectors p_n and q_n so as to be block A-biorthogonal. Here, similar to the V–W sequence, the vectors p_n and q_n are also constructed using look-ahead techniques. For example, we again have blocks $P^{(j)}$,

$$P^{(j)} = [\, p_{m_j} \quad p_{m_j+1} \quad \cdots \quad],$$

where p_{m_j} is called regular, the other vectors in the block are called inner, and the indices m_j satisfy

$$1 =: m_1 < m_2 < \cdots < m_k < n \le m_{k+1}, \quad k := k(n).$$

The regular vectors p_{m_j} satisfy the A-biorthogonality condition

$$q_i^T A p_{m_j} = 0 \quad \text{for all} \quad i < m_j, \tag{3.2}$$

while the inner vectors satisfy only a relaxed version of this condition. Once again, the structure of Q_n parallels that of P_n, and the A-biorthogonality of the two sets of basis vectors can be written as:

$$Q_n^T A P_n = E_n = \operatorname{diag}(E^{(1)}, E^{(2)}, \ldots, E^{(k)}), \quad E^{(j)} := (Q^{(j)})^T A P^{(j)}. \tag{3.3}$$

Before we consider the implementation details, let us briefly discuss an outline of the algorithm. At each iteration, the process consists of the following four basic steps.

Algorithm 3.1. (Overview of the coupled algorithm with look-ahead)

1) *Decide whether to construct* p_n *and* q_n *as regular or inner vectors.*
2) *Compute* p_n *and* q_n *as either regular or inner vectors.*
3) *Decide whether to construct* v_{n+1} *and* w_{n+1} *as regular or inner vectors.*
4) *Compute* v_{n+1} *and* w_{n+1} *as either regular or inner vectors.*

Steps 1) and 3) are the basis of the look-ahead strategy, and they each consist of three checks. Recall from (2.6) and (2.7) that the Lanczos vectors v_{n+1} and w_{n+1} can be obtained from the previous Lanczos vectors by a block three-term recurrence. Similarly, it is possible to show that the vectors p_n and q_n also have a block three-term recurrence, of the form

$$AP_{n-1} = P_n G_{n-1} \quad \text{and} \quad A^T Q_{n-1} = Q_n \Gamma_n^{-1} G_{n-1} \Gamma_{n-1}, \tag{3.4}$$

where

$$G_{n-1} := U_n L_{n-1}. \tag{3.5}$$

The look-ahead strategy for the two pairs of sequences is then similar, and was first proposed in [3]. In Step 1), the algorithm checks whether:

$$\sigma_{\min}(E^{(k)}) \geq eps,$$

$$n(A)\|p_n\| \geq \sum_{i=m_{k-1}}^{n-1} \left|(U_n L_{n-1})_{i,n-1}\right| \|p_i\|, \tag{3.6}$$

$$n(A)\|q_n\| \geq \sum_{i=m_{k-1}}^{n-1} \frac{\gamma_{n-1}}{\gamma_i} \left|(U_n L_{n-1})_{i,n-1}\right| \|q_i\|. \tag{3.7}$$

Here, *eps* is machine epsilon, and $n(A)$ is an estimate for the norm of A. The vectors p_n and q_n are built as regular vectors only if all three of the above checks are satisfied. Likewise, in Step 3), the algorithm checks whether:

$$\sigma_{\min}(D^{(l)}) \geq eps,$$

$$n(A) \geq \sum_{i=n_{l-1}}^{n} |(L_n U_n)_{in}|, \tag{3.8}$$

$$n(A) \geq \sum_{i=n_{l-1}}^{n} \frac{\gamma_n}{\gamma_i} |(L_n U_n)_{in}|, \tag{3.9}$$

and again, the vectors v_{n+1} and w_{n+1} are built as regular vectors only if all three of the checks are passed. The motivation for these checks can be found in [3, 6]. Here, we will only note that the look-ahead strategy proposed will build regular vectors in preference to inner vectors, and thus it will take as few look-ahead steps as possible.

Once the decisions in Steps 1) and 3) are taken, the next vectors p_n and q_n, and v_{n+1} and w_{n+1}, are built in Steps 2) and 4). For p_n and q_n, let $k^\star$ denote the number of the row of the first possible nonzero element in the nth column of U_n. It can be shown that

$$k^\star = \max\{j \mid 1 \le j \le k \text{ and } m_j \le \max\{1, n_l - 1\}\}. \tag{3.10}$$

From (3.3), both the regular and the inner vectors have the same coefficients $U_{m_{k^\star}:m_k-1,n}$, given by[†]

$$U_{m_{k^\star}:m_k-1,n} = (\operatorname{diag}(E^{(k^\star)}, E^{(k^\star+1)}, \ldots, E^{(k-1)}))^{-1} \cdot [\,Q^{(k^\star)} \quad Q^{(k^\star+1)} \quad \cdots \quad Q^{(k-1)}\,]^T A v_n.$$

For the regular vectors, the coefficients $U_{m_k:n-1,n}$ are determined by the condition (3.2),

$$U_{m_k:n-1,n} = (E^{(k)})^{-1}(Q^{(k)})^T A v_n, \tag{3.11}$$

while for the inner vectors, the coefficients $U_{m_k:n-1,n}$ are arbitrary:

$$u_{in} = \zeta_{in} \in \mathbb{C}, \quad \text{for} \quad i = m_k, m_k + 1, \ldots, n-1. \tag{3.12}$$

This completes the computation of the recurrence coefficients for p_n and q_n, and the vectors are then computed from

$$p_n = v_n - \sum_{i=m_{k^\star}}^{n-1} p_i u_{in},$$

$$q_n = w_n - \sum_{i=m_{k^\star}}^{n-1} q_i u_{in} (\gamma_n/\gamma_i).$$

For v_{n+1} and w_{n+1}, let $l^\star$ denote the number of the row of the first possible nonzero element in the nth column of L_n. It can be shown that

$$l^\star = \max\{j \mid 1 \le j \le l \text{ and } n_j \le m_k\}. \tag{3.13}$$

Again, by (2.3), both the regular and the inner vectors have the same coefficients $L_{n_{l^\star}:n_l-1,n}$, given by

$$L_{n_{l^\star}:n_l-1,n} = (\operatorname{diag}(D^{(l^\star)}, D^{(l^\star+1)}, \ldots, D^{(l-1)}))^{-1} \cdot [\,W^{(l^\star)} \quad W^{(l^\star+1)} \quad \cdots \quad W^{(l-1)}\,]^T A p_n.$$

For the regular vectors, the coefficients $L_{n_l:n,n}$ are determined by the condition (2.2),

$$L_{n_l:n,n} = (D^{(l)})^{-1}(W^{(l)})^T A p_n, \tag{3.14}$$

while for the inner vectors, the coefficients $L_{n_l:n,n}$ are arbitrary:

$$l_{in} = \eta_{in} \in \mathbb{C} \quad \text{for} \quad i = n_l, n_l + 1, \ldots, n. \tag{3.15}$$

[†]We denote $x_{i:j} = [\,x_i \quad x_{i+1} \quad \cdots \quad x_j\,]^T$.

Once the recurrence coefficients for v_{n+1} and w_{n+1} are computed, the vectors are constructed by scaling the vectors

$$\tilde{v}_{n+1} = Ap_n - \sum_{i=n_{l^\star}}^{n} v_i l_{in},$$

$$\tilde{w}_{n+1} = A^T q_n - \sum_{i=n_{l^\star}}^{n} w_i l_{in}(\gamma_n/\gamma_i),$$

to have unit length.

4 Implementation details

We now turn to a detailed description of the implementation of Algorithm 3.1. If the coupled two-term Lanczos process is run without any look-ahead steps, it will require two inner products at each step in order to compute all the coefficients of the recurrence formulas. Hence, the goal for the look-ahead implementation is to also require only two inner products per step for all the recurrence coefficients. Recall that the look-ahead strategy (3.6) and (3.7) for the P–Q sequence and the normalization (2.5) require a total of four norm computations, so that the implementation will require two inner products and four norms per iteration.

To begin, let us introduce the auxiliary matrices

$$F_n := W_n^T A P_n \quad \text{and} \quad \tilde{F}_n := Q_n^T A V_n,$$

whose columns are needed in (3.14) and in (3.11). In addition, we will make use of the following symmetry relations from [6]:

$$D_n \Gamma_n = (D_n \Gamma_n)^T, \tag{4.1}$$

$$E_n \Gamma_n = (E_n \Gamma_n)^T, \tag{4.2}$$

$$F_n \Gamma_n = (\tilde{F}_n \Gamma_n)^T. \tag{4.3}$$

The nth iteration of the implementation will update the matrices D_{n-1}, E_{n-1}, F_{n-1}, L_{n-1}, U_{n-1}, P_{n-1}, Q_{n-1}, V_n, and W_n, to D_n, E_n, F_n, L_n, U_n, P_n, Q_n, V_{n+1}, and W_{n+1}, respectively. We first list an outline of the algorithm as we have implemented it.

Algorithm 4.1. (Coupled algorithm with look-ahead)

0) *Choose* $v_1, w_1 \in \mathbb{C}^N$ *with* $\|v_1\| = \|w_1\| = 1$, *and compute* $w_1^T v_1$.
   Set $k = 1$, $m_k = 1$, $l = 1$, $n_l = 1$.

For $n = 1, 2, \ldots,$ *do*:

1) *Update* D_{n-1} *to* D_n.
2) *Determine* $k^\star$ *from* (3.10):

$$k^\star = \max\left\{ j \mid 1 \le j \le k \text{ and } m_j \le \max\{1, n_l - 1\} \right\}.$$

3) *Compute* $F_{n,1:n-1}$ *from* (4.4) *below, using* L_{n-1} *and* $D_{n,1:n}$.
Then compute $\tilde{F}_{1:n-1,n}$ *from* (4.3).

4) *Check whether* $E^{(k)}$ *is nonsingular*:

$$\texttt{innerp} = \sigma_{\min}(E^{(k)}) < eps.$$

5) *Compute the part of* $U_{1:n,n}$ *that is determined by biorthogonality*:

$$\begin{aligned} U_{m_i:m_{i+1}-1,n} &= (E^{(i)})^{-1}(Q^{(i)})^T A v_n \\ &= (E^{(i)})^{-1}\tilde{F}_{m_i:m_{i+1}-1,n}, \quad i = k^\star, \dots, k-1. \end{aligned}$$

If **innerp**, *go to* 6). *Otherwise, set*

$$U_{m_k:n-1,n} = (E^{(k)})^{-1}(Q^{(k)})^T A v_n = (E^{(k)})^{-1}\tilde{F}_{m_k:n-1,n}.$$

6) *Build the part of* p_n *and* q_n *that is common to both regular and inner vectors*:

$$p_n = v_n - \sum_{i=m_{k^\star}}^{m_k-1} p_i u_{in},$$

$$q_n = w_n - \sum_{i=m_{k^\star}}^{m_k-1} q_i u_{in}(\gamma_n/\gamma_i).$$

If **innerp**, *go to* 11).

7) *Build* $G_{m_k:n-1,n-1}$ *and check the coefficient* $G_{m_{k-1}:n-1,n-1}$.
If **innerp**, *go to* 11).

8) *Build* p_n *and* q_n *as regular vectors*:

$$p_n = p_n - \sum_{i=m_k}^{n-1} p_i u_{in},$$

$$q_n = q_n - \sum_{i=m_k}^{n-1} q_i u_{in}(\gamma_n/\gamma_i).$$

Compute Ap_n, $q_n^T A p_n$, $\|p_n\|$, *and* $\|q_n\|$.

9) *Build and check the coefficient* $G_{m_k:n-1,n}$. *If* **innerp**, *go to* 11).

10) *Set* $m_{k+1} = n$, $k = k+1$, *and go to* 12).

11) *Choose the inner recurrence coefficients* u_{in}, $i = m_k, \dots, n-1$, *and build* p_n *and* q_n *as inner vectors*:

$$p_n = p_n - \sum_{i=m_k}^{n-1} p_i u_{in},$$

$$q_n = q_n - \sum_{i=m_k}^{n-1} q_i u_{in}(\gamma_n/\gamma_i).$$

Compute Ap_n, $q_n^T Ap_n$, $\|p_n\|$, *and* $\|q_n\|$.

12) *If* $\|p_n\| = 0$, *or* $\|q_n\| = 0$, *then stop.*

13) *Compute* $A^T q_n$.

14) *Update* E_{n-1} *to* E_n.

15) *Determine* $l^\star$ *from* (3.13):

$$l^\star = \max\left\{ j \mid 1 \leq j \leq l \text{ and } n_j \leq m_k \right\}.$$

16) *Compute* $F_{1:n,n}$ *from* (4.5) *below, using* E_n *and* U_n.

17) *Check whether* $D^{(l)}$ *is nonsingular*:

$$\texttt{innerv} = \sigma_{\min}(D^{(l)}) < eps.$$

18) *Compute the part of* $L_{1:n,n}$ *that is determined by biorthogonality*:

$$\begin{aligned} L_{n_i:n_{i+1}-1,n} &= (D^{(i)})^{-1}(W^{(i)})^T Ap_n \\ &= (D^{(i)})^{-1} F_{n_i:n_{i+1}-1,n}, \quad i = l^\star, \ldots, l-1. \end{aligned}$$

If `innerv`, *go to* 19). *Otherwise, set*

$$L_{n_l:n,n} = (D^{(l)})^{-1}(W^{(l)})^T Ap_n = (D^{(l)})^{-1} F_{n_l:n,n}.$$

19) *Build the part of* v_{n+1} *and* w_{n+1} *that is common to both regular and inner vectors*:

$$\tilde{v}_{n+1} = Ap_n - \sum_{i=n_{l^\star}}^{n_l-1} v_i l_{in},$$

$$\tilde{w}_{n+1} = A^T q_n - \sum_{i=n_{l^\star}}^{n_l-1} w_i l_{in} (\gamma_n/\gamma_i).$$

If `innerv`, *go to* 24).

20) *Build* $H_{n_l:n,n}$ *and check the coefficient* $H_{n_{l-1}:n,n}$.

If `innerv`, *go to* 24).

21) *Build* v_{n+1} *and* w_{n+1} *as regular vectors*:

$$\tilde{v}_{n+1} = \tilde{v}_{n+1} - \sum_{i=n_l}^{n} v_i l_{in},$$

$$\tilde{w}_{n+1} = \tilde{w}_{n+1} - \sum_{i=n_l}^{n} w_i l_{in} (\gamma_n/\gamma_i).$$

Compute $\rho_{n+1} = l_{n+1,n} = \|\tilde{v}_{n+1}\|$, $\xi_{n+1} = \|\tilde{w}_{n+1}\|$.

If $\rho_{n+1} = 0$ *or* $\xi_{n+1} = 0$, *then stop.*

Otherwise, set $\gamma_{n+1} = \gamma_n \rho_{n+1}/\xi_{n+1}$, *and compute* $\tilde{w}_{n+1}^T \tilde{v}_{n+1}$.

22) *Build and check the coefficient* $H_{n_l:n,n+1}$. *If* `innerv`, *go to* 24).

23) *Set* $n_{l+1} = n+1$, $l = l+1$, *and go to* 25).

24) *Choose the inner recurrence coefficients* l_{in}, $i = n_l, \ldots, n$, *and build* v_{n+1} *and* w_{n+1} *as inner vectors:*

$$\tilde{v}_{n+1} = \tilde{v}_{n+1} - \sum_{i=n_l}^{n} v_i l_{in},$$

$$\tilde{w}_{n+1} = \tilde{w}_{n+1} - \sum_{i=n_l}^{n} w_i l_{in}(\gamma_n/\gamma_i).$$

Compute $\rho_{n+1} = l_{n+1,n} = \|\tilde{v}_{n+1}\|$, $\xi_{n+1} = \|\tilde{w}_{n+1}\|$.
If $\rho_{n+1} = 0$ *or* $\xi_{n+1} = 0$, *then stop.*
Otherwise, set $\gamma_{n+1} = \gamma_n \rho_{n+1}/\xi_{n+1}$, *and compute* $\tilde{w}_{n+1}^T \tilde{v}_{n+1}$.

25) *Set*

$$v_{n+1} = \tilde{v}_{n+1}/\rho_{n+1}, \quad w_{n+1} = \tilde{w}_{n+1}/\xi_{n+1},$$
$$w_{n+1}^T v_{n+1} = \tilde{w}_{n+1}^T \tilde{v}_{n+1}/(\rho_{n+1}\xi_{n+1}).$$

Step 1. The diagonal term $w_n^T v_n$ has already been computed directly, at the end of the previous step. Next, using

$$\begin{aligned} F_{n-1} &= W_{n-1}^T A P_{n-1} = W_{n-1}^T V_n L_{n-1} \\ &= D_{n-1} L_{1:n-1,1:n-1} + l_{n,n-1} D_{1:n-1,n} \begin{bmatrix} 0 & \cdots & 0 & 1 \end{bmatrix}, \end{aligned} \tag{4.4}$$

the remainder of the last column of D_n is computed from D_{n-1}, F_{n-1}, and L_{n-1}. The last row of D_n is obtained by symmetry, using (4.1).

Step 7. We build $G_{m_k:n-1,n-1}$, which would be the coefficient of the $P^{(k)}$ and $Q^{(k)}$ blocks in the three-term recurrences (3.4) for p_n and q_n. Using (3.5), one has

$$G_{i,n-1} = \sum_{j=i}^{n} u_{ij} l_{j,n-1}, \quad i = m_k, \ldots, n-1.$$

The coefficient $G_{m_{k-1}:m_k-1,n-1}$ has already been built as part of Step 9) at the previous iteration. We then check (3.6)–(3.7), and set `innerp` to `TRUE` if at least one of the two checks fails.

Step 9. We build $G_{m_k:n-1,n}$, which would be the coefficient of the $P^{(k)}$ and $Q^{(k)}$ blocks in the three-term recurrences (3.4) for p_{n+1} and q_{n+1}. It is straightforward to show that

$$G_{m_k:n-1,n} = (E^{(k)})^{-1} (Q^{(k)})^T A A p_n.$$

Moreover, we have

$$\begin{aligned} Q_{n-1}^T A A p_n &= \left(A^T Q_{n-1}\right)^T A p_n = \left(Q_n \Gamma_n^{-1} U_n L_{n-1} \Gamma_{n-1}\right)^T A p_n \\ &= \Gamma_{n-1} L_{n-1}^T U_n^T \Gamma_n^{-1} Q_n^T A p_n \\ &= \Gamma_{n-1} L_{n-1}^T U_n^T \Gamma_n^{-1} \begin{bmatrix} 0 & \cdots & 0 & q_n^T A p_n \end{bmatrix}^T \\ &= \frac{\gamma_{n-1}}{\gamma_n} l_{n,n-1} \begin{bmatrix} 0 & \cdots & 0 & q_n^T A p_n \end{bmatrix}^T . \end{aligned}$$

We then check a subset of (3.6)–(3.7), and set `innerp` to `TRUE` if at least one of the two checks fails.

Step 14. The diagonal term $q_n^T A p_n$ has already been computed directly, as part of Step 8) or Step 11). Next, using

$$F_n = W_n^T A P_n = \Gamma_n U_n^T \Gamma_n^{-1} Q_n^T A P_n, \tag{4.5}$$

the remainder of the last row of E_n is computed from E_{n-1}, $F_{1:n,1:n-1}$, and U_n. The last column of E_n is obtained by symmetry, using (4.2).

Step 20. We build $H_{n_l:n,n}$, which would be the coefficient of the $V^{(l)}$ and $W^{(l)}$ blocks in the three-term recurrences (2.6) and (2.7) for v_{n+1} and w_{n+1}. Using (3.1), one has

$$H_{in} = \sum_{j=i}^{n} l_{ij} u_{jn}, \quad i = n_l, \ldots, n.$$

The coefficient $H_{n_{l-1}:n_l-1,n}$ has already been built as part of Step 22) at the previous iteration. We then check (3.8)–(3.9), and set `innerv` to `TRUE` if at least one of the two checks fails.

Step 22. We build $H_{n_l:n,n+1}$, which would be the coefficient of the $V^{(l)}$ and $W^{(l)}$ blocks in the three-term recurrences (2.6) and (2.7) for v_{n+2} and w_{n+2}. It is straightforward to show that

$$H_{n_l:n,n+1} = (D^{(l)})^{-1} (W^{(l)})^T A v_{n+1}.$$

Moreover, we have

$$\begin{aligned} W_n^T A v_{n+1} &= \left(A^T W_n\right)^T A v_{n+1} = \left(W_{n+1} \Gamma_{n+1}^{-1} L_n U_n \Gamma_n\right)^T v_{n+1} \\ &= \Gamma_n U_n^T L_n^T \Gamma_{n+1}^{-1} W_{n+1}^T v_{n+1} \\ &= \Gamma_n U_n^T L_n^T \Gamma_{n+1}^{-1} \begin{bmatrix} 0 & \cdots & 0 & w_{n+1}^T v_{n+1} \end{bmatrix}^T \\ &= \frac{\gamma_n}{\gamma_{n+1}} l_{n+1,n} \begin{bmatrix} 0 & \cdots & 0 & w_{n+1}^T v_{n+1} \end{bmatrix}^T . \end{aligned}$$

We then check a subset of (3.8)–(3.9), and set `innerv` to `TRUE` if at least one of the two checks fails.

We remark that the checks in steps 9) and 22) are actually slightly relaxed versions of (3.8)–(3.9), and (3.6)–(3.7), respectively, since the indices checked are only a subset of the full range appearing in (3.8)–(3.9) and (3.6)–(3.7). We also note that the algorithm above requires minimal inputs from the user. Recall that *eps* in steps 4) and 17) is machine epsilon. Furthermore, the estimate $n(A)$ for the norm of the matrix can be updated dynamically, as was done in [3].

The coupled Lanczos Algorithm 4.1 requires per iteration the computation of two inner products and four vector norms. We conclude this section by noting that, in Algorithm 4.1, the choice of the inner recurrence coefficients (3.12) and (3.15) is arbitrary. In our implementation of the algorithm, we have used

$$\begin{aligned}
u_{n-1,n} &= 1, \\
u_{n-2,n} &= 1, \quad \text{when} \quad m_k \leq n-2, \\
u_{in} &= 0, \quad \text{for} \quad i = m_k, \ldots, n-3, \\
l_{nn} &= 1, \\
l_{n-1,n} &= 1, \quad \text{when} \quad n_l \leq n-1, \\
l_{in} &= 0, \quad \text{for} \quad i = n_l, \ldots, n-2,
\end{aligned}$$

for the inner vector recurrence coefficients.

5 The coupled two-term QMR algorithm

We now consider the quasi-minimal residual approach and briefly outline how it can be combined with the coupled two-term look-ahead Lanczos algorithm of Section 3 to obtain a new implementation of the QMR method. We note that the QMR algorithm was proposed in [5] for the case of nonsingular linear systems (1.1). It was later shown by Freund and Hochbruck [4] that the algorithm can also be applied to singular systems, and that it always generates well-defined iterates. However, these iterates converge to a meaningful solution only for the special case of consistent systems with coefficient matrices A of index 1. Here, we consider the QMR method for the general case of $N \times N$ linear systems, with singular or nonsingular coefficient matrices.

The QMR algorithm belongs to the family of Krylov subspace methods. Let $x_0 \in \mathbb{C}^N$ be an initial guess for the solution of (1.1), and $r_0 = b - Ax_0$ the corresponding initial residual, of length $\rho_1 = \|r_0\|$. Choosing $v_1 = r_0/\rho_1$ as the starting right Lanczos vector for Algorithm 4.1, and w_1 with $\|w_1\| = 1$ as an arbitrary starting left Lanczos vector, one obtains the four basis sets, V_n, W_n, P_n, and Q_n, of which the ones of interest are V_n and P_n, related by:

$$V_n = P_n R_n \quad \text{and} \quad AP_n = V_{n+1} L_n.$$

Once the basis vectors are constructed, the nth QMR iterate is selected from the shifted Krylov subspace $x_0 + K_n(r_0, A)$ as

$$x_n = x_0 + P_n y_n, \tag{5.1}$$

where $y_n \in \mathbb{C}^n$ is defined by the quasi-minimal residual condition

$$\|f_{n+1} - L_n y_n\| = \min_{y \in \mathbb{C}^n} \|f_{n+1} - L_n y\|. \tag{5.2}$$

This is an $(n+1) \times n$ least-squares problem, where

$$f_{n+1} := \rho_1 \cdot [1 \quad 0 \quad \cdots \quad 0]^T \in \mathbb{R}^{n+1},$$

and we have used the normalization (2.5) of the Lanczos vectors; otherwise, the least-squares problem above also involves a diagonal scaling matrix.

Note that, by setting

$$z_n = (R_n)^{-1} y_n,$$

and inserting in (5.2), one obtains the equivalent least-squares problem

$$\|f_{n+1} - L_n R_n z_n\| = \min_{z \in \mathbb{C}^n} \|f_{n+1} - L_n R_n z\|,$$

which is exactly the least-squares problem solved by the QMR algorithm based on the three-term Lanczos process. Thus, the QMR iterates (5.1) are, in exact arithmetic, identical to the iterates of the original QMR algorithm [5]. However, as was shown in [6], in finite precision arithmetic, the coupled QMR algorithm is more robust than the three-term recurrence version.

Like the original QMR algorithm, the new implementation has a number of desirable properties. Since they are equivalent, all the theoretical properties of the three-term recurrence QMR algorithm carry over to the coupled version. The new method also shows a smooth and nearly monotone convergence curve. At all times during the iteration, an upper bound for the QMR residual norm is available at no extra cost, and, as the examples will show, this upper bound is a very good indicator of the convergence of the method. As in the original version, the QMR iterate has a short update formula, involving only one direction vector that can be updated with a two-term recurrence. This is slightly cheaper than in the old method, where the corresponding search directions for the update of the iterates had a three-term recurrence. Finally, the new version seems to be significantly more robust than the old version, though no theoretical results are available to explain the differences. For a full discussion of the properties of the two QMR algorithms, see [5, 6].

6 Numerical experiments

In this section, we present a few numerical examples with the new implementation of the QMR algorithm. All these examples were run either on a Cray–2 at the NASA Ames Research Center or on a Cray Y–MP at AT&T Bell Laboratories, with a machine epsilon of about 5.0E−29.

In the plots below, we always show two curves, the computed scaled residual norm $\|r_n\|/\|r_0\|$ (solid line) and the residual norm upper bound (dotted line). Recall that the upper bound is available at each step at no extra cost.

Example 6.1. This example is taken from [2]. Here we consider the differential equation

$$Lu = f \quad \text{on} \quad (0,1) \times (0,1) \times (0,1), \tag{6.1}$$

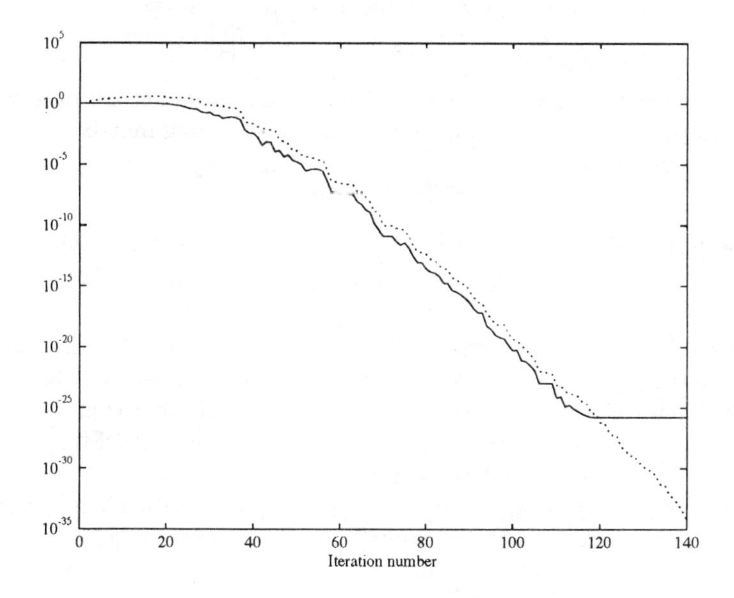

Figure 1: Convergence curves for Example 6.1.

where

$$Lu := -\frac{\partial}{\partial x}\left(e^{-xy}\frac{\partial u}{\partial x}\right) - \frac{\partial}{\partial y}\left(e^{xy}\frac{\partial u}{\partial y}\right) - \frac{\partial}{\partial z}\left(e^{xy}\frac{\partial u}{\partial z}\right)$$
$$+50(x+y+z)\frac{\partial u}{\partial x} + \left(\frac{1}{1+x+y+z} - 250\right)u,$$

with Dirichlet boundary conditions $u = 0$. The right-hand side f was chosen so that

$$u = (1-x)(1-y)(1-z)(1-e^{-x})(1-e^{-y})(1-e^{-z})$$

is the exact solution of (6.1). We discretized (6.1) using centered differences on a $40 \times 40 \times 40$ grid with mesh size $h = 1/41$. This leads to a linear system of order $N = 64000$ with 438400 nonzero entries. The starting vector w_1 was a random vector, and the initial guess x_0 was zero. The example was run with a right SSOR preconditioner [1], with $\omega = 1.0$. The algorithm stagnated at 1.6E−25 after 119 steps. For the V–W sequence, it built 4 blocks of size 2, and forced a block closure once. For the P–Q sequence, it built 2 blocks of size 2, and forced a block closure once.

Example 6.2. This example was provided to us by V. Venkatakrishnan [10], from the Numerical Aerodynamic Simulation Group at the NASA Ames Research Center. It comes from an unstructured 2-D Euler solver, and it corresponds to the system at the beginning of time-stepping. The linear system is of order $N = 62424$ with 1717792 nonzero elements. The right-hand side b and the starting vector w_1

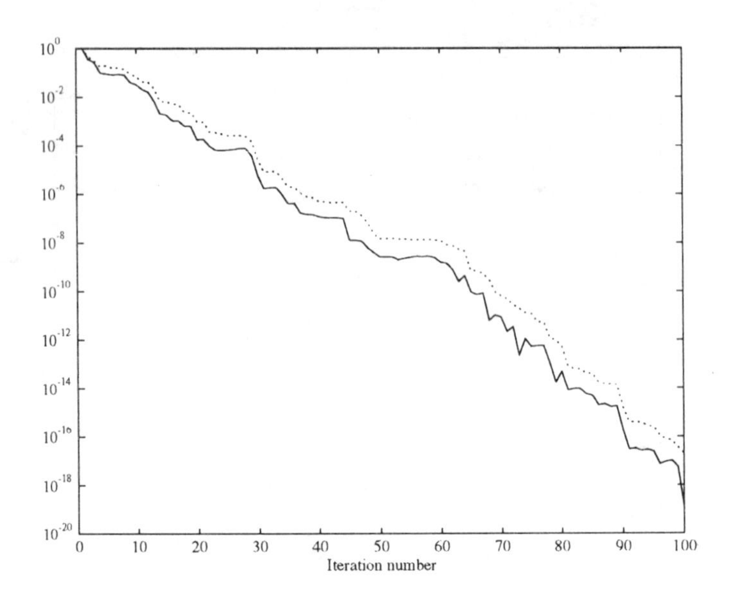

Figure 2: Convergence curves for Example 6.2.

were both random vectors, while the initial guess x_0 was zero. Once again, the example was run with a right SSOR preconditioner, with $\omega = 1.0$. The algorithm was stopped once it reached 1.0E−20, after 107 steps. For the V–W sequence, it built 3 blocks of size 2, and 1 block of size 3. For the P–Q sequence, it built no blocks, but forced a block closure once.

Example 6.3. This example was taken from [8], where McQuain and his collaborators investigated the applicability of iterative methods to linear systems arising in circuit simulation. In particular, they studied the solution of systems involving the Jacobians that arise when a homotopy algorithm is applied to the computation of the DC operating point of a circuit. While these linear systems seem to be rather difficult, they are not intractable. In the example, a Jacobian of order $N = 1853$ with 8994 nonzero elements was considered; it is the first Jacobian from the IS7B sequence discussed in Section 5.3 of [8]. The right-hand side b was obtained by moving the last column of the original rectangular $n \times (n+1)$ Jacobian to the other side. The starting guess x_0 was zero, and the starting vector w_1 was set $w_1 = v_1 = b$. The example was run with the variant described in [5] of Saad's ILUT preconditioner [9], with no additional fill-in allowed and a drop tolerance of 0.001, which generated a matrix L with 4766 nonzero elements, and a matrix U with 5034 nonzero elements. The algorithm stagnated at 8.7E−23 after 67 steps. It built no look-ahead blocks.

The examples shown illustrate several points. As already noted, the coupled QMR algorithm has a rather smooth and almost monotone convergence behavior.

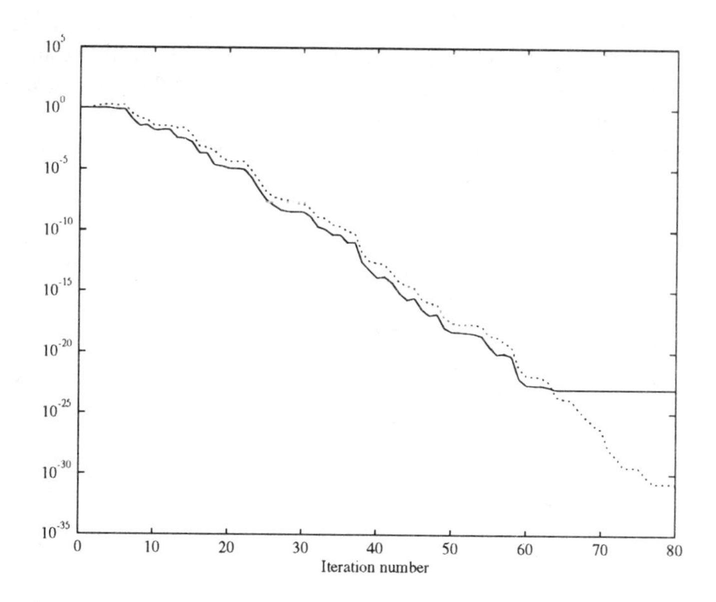

Figure 3: Convergence curves for Example 6.3.

It makes available a residual norm upper bound that is a very good indicator of the convergence of the method. Finally, while it is possible to construct examples with arbitrary look-ahead structure, numerical experience seems to indicate that, on the average, the look-ahead strategy does not build many look-ahead blocks. Indeed, the vast majority of the steps taken by the coupled two-term look-ahead Lanczos process are regular steps; furthermore, from the look-ahead steps of size greater than 1 taken, the majority are blocks of size 2.

7 Concluding remarks

We have presented details of an implementation of a new look-ahead algorithm for constructing Lanczos vectors based on coupled two-term recurrences instead of the usual three-term recurrences. We then discussed a new implementation of the quasi-minimal residual algorithm, using the coupled process to build the basis for the Krylov subspace. While the theoretical results derived for the original algorithm carry over to the new one, the latter was shown in examples to have better numerical properties.

FORTRAN 77 codes for the coupled-two term look-ahead procedure and the resulting new implementation of the QMR algorithm can be obtained electronically from the authors (freund@research.att.com or na.nachtigal@na-net.ornl.gov).

We note that FORTRAN 77 codes for the original implementation of QMR and the underlying look-ahead Lanczos algorithm are available from netlib by sending an email message consisting of the single line "send lalqmr from misc" to netlib@ornl.gov or netlib@research.att.com.

Acknowledgements. We would like to thank V. Venkatakrishnan and W. McQuain for providing the matrices for Examples 6.2 and 6.3.

References

[1] O. AXELSSON, *A survey of preconditioned iterative methods for linear systems of algebraic equations*, BIT **25** (1985) 166–187.

[2] D. BAXTER, J. SALTZ, M. SCHULTZ, S. EISENSTAT, AND K. CROWLEY, *An experimental study of methods for parallel preconditioned Krylov methods*, Technical Report RR–629, Department of Computer Science, Yale University, New Haven, Connecticut, 1988.

[3] R.W. FREUND, M.H. GUTKNECHT, AND N.M. NACHTIGAL, *An implementation of the look-ahead Lanczos algorithm for non-Hermitian matrices*, SIAM J. Sci. Comput. **14** (1993), to appear.

[4] R.W. FREUND AND M. HOCHBRUCK, *On the use of two* QMR *algorithms for solving singular systems and applications in Markov chain modeling*, Journal of Numerical Linear Algebra with Applications **2** (1993), to appear.

[5] R.W. FREUND AND N.M. NACHTIGAL, QMR: *a quasi-minimal residual method for non-Hermitian linear systems*, Numer. Math. **60** (1991), 315–339.

[6] R.W. FREUND AND N.M. NACHTIGAL, *An implementation of the* QMR *method based on coupled two-term recurrences*, Technical Report 92.15, RIACS, NASA Ames Research Center, Moffett Field, California, June 1992.

[7] C. LANCZOS, *An iteration method for the solution of the eigenvalue problem of linear differential and integral operators*, J. Res. Nat. Bur. Standards **45** (1950), 255–282.

[8] W.D. MCQUAIN, C.J. RIBBENS, L.T. WATSON, AND R.C. MELVILLE, *Preconditioned iterative methods for sparse linear algebra problems arising in circuit simulation*, Technical Report 92–07, Department of Computer Science, Virginia Polytechnic Institute and State University, Blacksburg, Virginia, March 1992.

[9] Y. SAAD, ILUT*: a dual threshold incomplete* LU *factorization*, Research Report UMSI 92/38, University of Minnesota Supercomputer Institute, Minneapolis, Minnesota, March 1992.

[10] V. VENKATAKRISHNAN AND T.J. BARTH, *Unstructured grid solvers on the* iPSC/860, Parallel Computational Fluid Dynamics, Rutgers, New Jersey, May 1992.

Preconditioned iterative regularization for ill–posed problems

*Martin Hanke**, *James Nagy*† *and Robert Plemmons*‡

Abstract. A preconditioned iterative regularization scheme is proposed and tested for solving large scale structured linear systems $H\mathbf{f} = \mathbf{g}+\eta$, arising from the discretization of ill-posed inverse problems in the presence of noise. The case where H is a block Toeplitz matrix with Toeplitz blocks is considered. The preconditioned conjugate gradient method is applied, where H is approximated by a block circulant matrix with circulant blocks. This results in a fast 2-D FFT-based iterative scheme with applications, *e.g.*, in image restoration. Thus we precondition in the Fourier domain, while iterating in the spatial domain. Our main purpose is to show how the iterations can be effectively and efficiently regularized for solving ill-posed problems by using the spectral decomposition of the preconditioner.

Key words. Toeplitz-block system, FFT–based preconditioners, ill-posed inverse problems, conjugate gradients, least squares, regularization, deconvolution, image restoration.

AMS(MOS) subject classifications. 65F10, 65F15, 43E10.

1 Introduction

In many engineering applications (*e.g.*, image restoration [12]) it is often necessary to numerically solve large scale structured $N \times N$ dimensional linear systems

$$H\mathbf{f} = \bar{\mathbf{g}}, \tag{1}$$

arising from discretization of ill-posed inverse problems. For example, (1) may result from the discretization of a first kind integral operator equation. Numerical

*Institut für Praktische Mathematik, Universität Karlsruhe, Englerstrasse 2, W-7500, Karlsruhe, Germany. Research supported by the Deutsche Forschungsgemeinschaft (DFG).

†Department of Mathematics, Southern Methodist University, Dallas, TX 75275-0461.

‡Department of Mathematics and Computer Science, Wake Forest University, Box 7388, Winston-Salem, NC 27109. Cray Y-MP time was provided by a grant from the North Carolina Supercomputing Center. Research supported by the US Air Force under grant no. AFOSR–91–0163 and NSF grant no. CCR–92–01105.

methods for such problems can be highly ill-conditioned because the minimum singular value of H is close to zero. Moreover, for these ill-posed problems the matrix H is typically full. Very often, however, the H will have a Toeplitz, or Toeplitz related, block structure. An $n \times n$ matrix T is called a Toeplitz matrix if the entries of T satisfy $t_{i,j} = t_{i-j}$, *i.e.*, the entries of T are constant down each diagonal. If the entries of an $n \times n$ Toeplitz matrix C also satisfy $c_{n-j} = c_{-j}$, then C is called a circulant matrix.

In this paper we will consider solving (1) when the matrix H is an $m \times m$ block Toeplitz matrix, with each block being an $n \times n$ (point) Toeplitz matrix. Specifically, we will assume that

$$H = \begin{bmatrix} T_0 & T_{-1} & \cdots & T_{-m+1} \\ T_1 & T_0 & \cdots & T_{-m+2} \\ \vdots & \vdots & & \vdots \\ T_{m-1} & T_{m-2} & \cdots & T_0 \end{bmatrix}, \tag{2}$$

where each block T_k is an $n \times n$ point Toeplitz matrix. We call a matrix of this form a *block Toeplitz with Toeplitz blocks* matrix, and denote it by BTTB(m, n). Such matrices arise naturally from discretization of 2-D Fredholm integral equations of the first kind with displacement kernels. Since, in this case, the continuous problem is ill-posed (cf. Groetsch [8]), it follows then that the matrix H will be very ill-conditioned; we call (1) a *discrete ill-posed inverse problem.* As a consequence of the ill-posedness, the solution $\mathbf{f}$ will be very sensitive to perturbations in the right hand side. This is an important observation since in most applications $\mathbf{g}$ is either a recorded signal or image, or is obtained from measurements.

In general, what is known is not $\bar{\mathbf{g}} = H\mathbf{f}$, but

$$\mathbf{g} = H\mathbf{f} + \boldsymbol{\eta},$$

where $\boldsymbol{\eta}$ represents unknown (Gaussian) noise or measurement errors. Because of the above comments on the ill-conditioned nature of H, one cannot simply solve $H\mathbf{f} = \mathbf{g}$ and expect to compute an accurate approximation to $\mathbf{f}$. In particular, $\mathbf{g}$ can be dominated by noise at high frequencies. Thus if H is ill-conditioned, then the *noise term* will become amplified at the higher frequencies, since the power spectrum of the degraded image is typically highest at low frequencies and rolls off significantly for higher ones. Therefore the higher frequencies, which are usually of more interest to the user, are degraded the most in approximating $\mathbf{f}$.

Regularization is often used to alleviate the sensitivity of the inverse problem to the perturbations to the right hand side. Regularization has the effect of filtering out the noise at the expense of restricting the set of admissible solutions. Various regularization procedures include Tikhonov [8], truncated SVD [11], and iterative methods[9], especially conjugate gradient iterations [17]. In each case, one has the non-trivial problem of choosing a *regularization parameter*, which controls the degree of bias in the solution. For example, in conjugate gradients, the regularization parameter is the iteration number, *i.e.*, the number of iterations applied.

Here we consider using the preconditioned conjugate gradient algorithm as a regularization procedure. Moreover, we study the potential use of a *block circulant with circulant blocks* (BCCB) matrix C as a preconditioner. BCCB preconditioners have been shown to be effective in speeding the convergence of the conjugate gradient algorithm for solving well-conditioned BTTB systems of equations [2, 6]. In addition, the Fast Fourier Transform (FFT) can be used in the computations. These preconditioners were extended to least squares problems by R. Chan, Nagy and Plemmons [3, 4], where it is shown how the FFT–based computations can be applied to ill-posed problems when Tikhonov regularization is used.

We stress that to the best of our knowledge preconditioning effects have not yet been considered for iterative regularization methods; the topic is especially delicate, for either the preconditioner is as badly conditioned as is H, in which case each iterative step will already be unstable, or else C is a poor approximation of H, in which case the preconditioner will not be effective. We will show here how to adapt the BCCB preconditioner of [4, 6] to the context of discrete ill-posed equations arising from image restoration applications.

This paper is outlined as follows. In Section 2 we review the construction of BCCB preconditioners for BTTB matrices. Section 3 is concerned with the relationship between regularization and the singular value decomposition. In Section 4 we discuss regularization by iteration, with an emphasis on preconditioned conjugate gradient iterations. In Sections 5 and 6 we consider the problems of constructing a BCCB preconditioner for ill-conditioned BTTB matrices. Here we propose a particular way of adapting the BCCB preconditioner matrix using its FFT-based spectral decomposition. Finally, in Section 7 we provide some numerical results illustrating the features and the effectiveness of our scheme.

2 Preconditioned 2-D Problems

In this section we review conjugate gradient preconditioners for the iterative solution of some large-scale BTTB Toeplitz, *i.e.* 2-D, problems. Toeplitz related matrices arise in very many scientific applications, such as signal and image processing.

Traditional direct methods for solving point $n \times n$ Toeplitz systems of linear equations have $O(n^2)$ computational complexity; see *e.g.*, [7]. Fast direct Toeplitz solvers have complexity $O(n \log^2 n)$, but their numerical stability is not completely understood. In 1986, Gilbert Strang [16] addressed the question of whether iterative methods can compete with direct methods for solving symmetric positive definite Toeplitz systems. Strang proposed the use of circulant matrices to precondition conjugate gradient iterations for Toeplitz systems in order to accelerate the convergence. The reason why this approach is competitive with direct methods is clear. The use of circulant preconditioners for these Toeplitz problems allows the use of Fourier transforms throughout the computations, and these FFT-based iterations are not only numerically efficient, but also highly parallelizable.

Numerous articles have extended the Strang idea to more general point Toeplitz systems, and several types of circulant preconditioners have been suggested, *e.g.*,

[5]. Recently, FFT-based preconditioners have been proposed for solving BTTB problems by T. Chan and Olkin [6] and, for the least squares case, by R. Chan, Nagy and Plemmons [3, 4].

The conjugate gradient (CG) method is an iterative method of fundamental importance for solving systems of linear equations (1); *cf.* Golub and van Loan [7]. The classical CG method applies only to the case where H is symmetric positive definite. But if H is not symmetric positive definite one can still use the CG method; for example, one can consider the factored form of the normal equations

$$H^*(\mathbf{g} - H\mathbf{f}) = \mathbf{0},$$

where $*$ denotes the conjugate transpose. This variant of CG is generally called CGNR.

The convergence rate of the conjugate gradient algorithm depends on the singular values of the coefficient matrix H. If the singular values cluster around a fixed point, convergence will be rapid; *cf.* van der Sluis and van der Vorst [18]. Thus, to make the algorithm a useful iterative method, one usually *preconditions* the system. The preconditioned conjugate gradient algorithm, with preconditioner C, uses the conjugate gradient method to solve

$$HC^{-1}\mathbf{y} = \mathbf{g},$$

with $C\mathbf{f} = \mathbf{y}$.

This leads to the preconditioned conjugate gradient (PCGNR) algorithm for general linear systems and for least squares problems. The version of the PCGNR algorithm we use is given in [1] and can be stated as follows:

Algorithm 1. PCGNR. *Let* $\mathbf{f}^{(0)}$ *be an initial vector, and let* C *be a given preconditioner. This algorithm computes the solution,* $\mathbf{f}$*, to* $H\mathbf{f} = \mathbf{g}$.

- $\mathbf{r}^{(0)} = \mathbf{g} - H\mathbf{f}^{(0)}$
- $\mathbf{p}^{(0)} = \mathbf{s}^{(0)} = C^{-*}H^*\mathbf{r}^{(0)}$
- $\gamma_0 = \|\mathbf{s}^{(0)}\|_2^2$
- `for` $k = 0, 1, 2, \ldots$

$$\begin{aligned}
&\mathbf{q}^{(k)} = HC^{-1}\mathbf{p}^{(k)} \\
&\alpha_k = \gamma_k / \|\mathbf{q}^{(k)}\|_2^2 \\
&\mathbf{f}^{(k+1)} = \mathbf{f}^{(k)} + \alpha_k C^{-1}\mathbf{p}^{(k)} \\
&\mathbf{r}^{(k+1)} = \mathbf{r}^{(k)} - \alpha_k \mathbf{q}^{(k)} \\
&\mathbf{s}^{(k+1)} = C^{-*}H^*\mathbf{r}^{(k+1)} \\
&\gamma_{k+1} = \|\mathbf{s}^{(k+1)}\|_2^2 \\
&\beta_k = \gamma_{k+1}/\gamma_k \\
&\mathbf{p}^{(k+1)} = \mathbf{s}^{(k+1)} + \beta_k \mathbf{p}^{(k)}
\end{aligned}$$

We consider the T. Chan [5] circulant approximation for a square point Toeplitz matrix. His circulant approximation which is optimal with respect to the Frobenius norm, is defined for arbitrary square matrices A, and is constructed by averaging the entries along certain pairs of diagonals of A. In particular, for a given generic $n \times n$ matrix A, let $\mathcal{C}(A)$ be the $n \times n$ circulant approximation of A as defined in T. Chan [5]; *i.e.*, C is the minimizer of $F(X) = \|A - X\|_F$ over all circulant matrices X. For the special case where $A \equiv T$, where T is Toeplitz, we have

$$C = \mathcal{C}(A) = (c_{j-\ell})_{j,\ell},$$

where the (j,ℓ)th entry of C is given by the diagonal

$$c_k = \begin{cases} \dfrac{(n-k)t_k + kt_{k-n}}{n} & 0 \le k < n, \\ c_{n+k} & 0 < -k < n. \end{cases} \tag{3}$$

Throughout this paper we will use the T. Chan approximation $\mathcal{C}(\cdot)$ exclusively as a basic tool to construct our preconditioners for BTTB problems.

We now describe a BCCB preconditioner for BTTB, *i.e.*, 2-D, problems. It is based on second level approximations and will be called the *Level–2 preconditioner*. Our Level–2 preconditioner is based on circulant approximations at the Toeplitz-block level and again at a second, block Toeplitz level. The BTTB matrix H in (2) will be approximated by a second-level circulant preconditioner, denoted by $\mathcal{C}_2(H)$. The preconditioner is related to the one proposed by T. Chan and Olkin [6] for square Toeplitz-block matrices and extended to least squares problems by R. Chan, Nagy and Plemmons [4].

For the BTTB matrix H given by (2), we define a first-level circulant approximation by

$$\mathcal{C}_1(H) \equiv \begin{bmatrix} \mathcal{C}(T_0) & \mathcal{C}(T_{-1}) & \cdots & \mathcal{C}(T_{-m+1}) \\ \mathcal{C}(T_1) & \mathcal{C}(T_0) & \cdots & \mathcal{C}(T_{-m+2}) \\ \vdots & \vdots & \ddots & \vdots \\ \mathcal{C}(T_{m-1}) & \mathcal{C}(T_{m-2}) & \cdots & \mathcal{C}(T_0) \end{bmatrix}. \tag{4}$$

Now we apply another first-level circulant approximation $\tilde{\mathcal{C}_1}$ to $\mathcal{C}_1(H)$ to get the second-level circulant approximation $\mathcal{C}_2(H)$ that we want. The matrix $\tilde{\mathcal{C}_1}\{\mathcal{C}_1(H)\}$ is obtained by treating each block $\mathcal{C}(T_\mu)$ in $\mathcal{C}_1(H)$ as an entry of a matrix and then applying the formula (3) to this "point" matrix. More precisely, we form the block-circulant matrix

$$\tilde{\mathcal{C}_1}\{\mathcal{C}_1(H)\} = \begin{bmatrix} \tilde{C}_0 & \tilde{C}_{-1} & \cdots & \tilde{C}_{-m+1} \\ \tilde{C}_1 & \tilde{C}_0 & \cdots & \tilde{C}_{-m+2} \\ \vdots & \vdots & \ddots & \vdots \\ \tilde{C}_{m-1} & \tilde{C}_{m-2} & \cdots & \tilde{C}_0 \end{bmatrix}$$

where the $\tilde{C}_\mu$ are computed using (3). Since each $\mathcal{C}(T_\mu)$ is circulant, we see that the sums $\tilde{C}_\mu$ are circulant matrices. Moreover, by their definition, we see that $\tilde{C}_\mu = \tilde{C}_{\mu-m}, \quad 0 < \mu < m$. In particular, we see that $\tilde{\mathcal{C}_1}\{\mathcal{C}_1(H)\}$ is then a block-circulant matrix with circulant blocks of the form BCCB(m,n). We note further that $\mathcal{C}_2(H) = \tilde{\mathcal{C}_1}\{\mathcal{C}_1(H)\}$, as given in R. Chan and Jin [2, Theorem 3]. It is shown in [2] that $\mathcal{C}_2(H)$ is "optimal" in the sense that it minimizes $\|H - X\|_F$ over all BCCB matrices X. For our purposes, we need only find the first column of $\mathcal{C}_1(H)$, and then use that column to find the first column of the final BCCB matrix $\mathcal{C}_2(H)$.

We note that the eigenvalues of $\mathcal{C}_2(H)$ can be obtained by taking the 2-D FFT of the first column of $\mathcal{C}_2(H)$. Specifically, if $\mathbf{c}_2$ denotes the first column of $\mathcal{C}_2(H)$

and fft2d denotes the 2-D FFT operator, then the vector of eigenvalues of $\mathcal{C}_2(H)$ is given by $\texttt{fft2d}(\mathbf{c}_2)$.

The construction of our Level–2 BCCB preconditioner (*i.e.*, the eigenvalues of $C \equiv \mathcal{C}_2(H)$) for the matrix H in (2) is summarized in the following algorithm:

Algorithm 2. Construction of the circulant **Level–2** preconditioner for 2-D problems. *Let H be the BTTB(m,n) matrix given as in* (2). *This algorithm computes the eigenvalues of the BCCB preconditioner C, using the 2-D FFT, and stores them in the diagonal matrix Λ.*

Find the first column, $\mathbf{c}_1$, of $\mathcal{C}_1(H)$, using (3).

Find the first column, $\mathbf{c}_2$, of $C \equiv \mathcal{C}_2(H) = \tilde{\mathcal{C}_1}\{\mathcal{C}_1(H)\}$, using (3).

Find Λ from $\Lambda \mathbf{1} = \texttt{fft2d}(\mathbf{c}_2)$, using the 2-D FFT.

Since H is $mn \times mn$, the cost of forming the Level–2 BCCB preconditioner C (*i.e.*, its eigenvalues in Λ), for the case where the matrix H is BTTB(m,n) , is $O(mn(\log n + \log m))$ operations. For implementation in the PCGNR algorithm, we need only the eigenvalues of C together with the 2-D FFT. We observe that for any vector $\mathbf{y}$, $C^{-1}\mathbf{y}$ can be computed by the formula

$$C^{-1}\mathbf{y} = \texttt{ifft2d}[\Lambda^{-1}\texttt{fft2d}(\mathbf{y})]$$

in

$$O(mn(\log n + \log m)) \tag{5}$$

operations by using fft2d and its inverse, ifft2d. FFT-based matrix-vector multiplications involving the BTTB matrices H and H^* in the PCGNR algorithm also have complexity (5), so (5) is the complexity per iteration of the PCGNR algorithm, implemented using the 2-D FFT.

We note that a convergence analysis of the PCGNR method with the BCCB preconditioner C is given in [4]. There it is shown that, for certain important classes of H, the singular values of the preconditioned matrix are clustered around one, leading to rapid convergence. It also follows from the analysis in [4] that the PCGNR iterations may be ineffective or even break down if H is very ill-conditioned. In the remaining sections of the paper, our purpose is to show how the iterations can be effectively and efficiently regularized for solving ill-posed problems by using the FFT-based spectral decomposition of C.

3 Regularization and the SVD

Let

$$H = U\,\Sigma\,V^T$$

be the singular value decomposition (SVD) of H, (cf. [7]) where the columns $\mathbf{u}_1, \ldots, \mathbf{u}_N$ of U and the columns $\mathbf{v}_1, \ldots, \mathbf{v}_N$ of V form orthonormal bases of $\mathbf{R}^N$, respectively, and Σ is a diagonal matrix containing the singular values $\sigma_1 \geq \sigma_2 \geq \ldots \geq \sigma_N > 0$ (we assume that H is nonsingular).

In applications arising from integral equations of the first kind, σ_1 is a moderately sized number – more or less independent of the discretization – and the ill-conditioning of H stems from clustering of the smaller singular values at the origin. In convolution problems over $\mathbf{R}^N$, such as those we will consider, the singular values typically fill out the entire interval $(0, \sigma_1]$ with no particular gap in their spectrum.

Expanding $\mathbf{f}$ and $\boldsymbol{\eta}$ in terms of the SVD we have

$$\mathbf{f} = \sum_{i=1}^{N} f_i \mathbf{v}_i, \qquad \boldsymbol{\eta} = \sum_{i=1}^{N} \eta_i \mathbf{u}_i,$$

which gives

$$\mathbf{g} = \sum_{i=1}^{N} (\mathbf{g}^T \mathbf{u}_i)\mathbf{u}_i = \sum_{i=1}^{N} (\sigma_i f_i + \eta_i)\mathbf{u}_i.$$

It is not our aim to rigorously treat the theory of regularization; for this we rather refer to the forthcoming survey [10]. Instead, we prefer to derive an understanding of the preconditioning issues in CGNR–based regularization. With this in mind, a heuristic treatment of regularization seems to be more appropriate.

Our discussion is based on the following two main assumptions:

- The spectral coefficients $\bar{\mathbf{g}}^T \mathbf{u}_i = \sigma_i f_i$ of the unperturbed right–hand side $\bar{\mathbf{g}} = H\mathbf{f}$ decay in absolute value like the σ_i; indeed, we always have

  $$|\bar{\mathbf{g}}^T \mathbf{u}_i| = \sigma_i |f_i| \leq \sigma_i ||\mathbf{f}||, \quad i = 1, 2, \ldots.$$

  This assumption reflects the *Picard condition* of the continuous problem (*cf.* [8]) which is even somewhat stronger. Following Hansen [11, Def. 3.3] we call this the *discrete Picard condition.*

- $\boldsymbol{\eta}$ originates from "Gaussian white noise"; that means its spectral components η_i are independent stochastic variables with mean zero and variance $\varepsilon^2 > 0$. We may therefore expect that

  $$|\eta_i| \approx \varepsilon, \qquad i = 1, 2, \ldots.$$

For practical reasons we will assume henceforth that the signal $\bar{\mathbf{g}}$ dominates the noise $\boldsymbol{\eta}$; no reasonable reconstructions are possible otherwise.

We separate the indices $i \in \{1, \ldots, N\}$ into three subsets: $\mathcal{I}_s$, corresponding to the *signal subspace* $\mathcal{U}_s := \text{span}\{\mathbf{u}_i : \varepsilon \ll \sigma_i\}$; $\mathcal{I}_n$, corresponding to the *noise subspace* $\mathcal{U}_n := \text{span}\{\mathbf{u}_i : \sigma \ll \varepsilon\}$; and the remaining ones $\mathcal{I}_t$, defining $\mathcal{U}_t := \text{span}\{\mathbf{u}_i : i \in \mathcal{I}_t\}$ as the *transition subspace*, where σ_i and ε have the same order of magnitude. While the first two sets constitute a highly simplifying distinction between "good" and "bad" singular values, the idea of $\mathcal{I}_t$ associated with the transition subspace is to cope with the inherent shortcomings of such a procedure.

Reconstructing naively, *i.e.*, without regularization, yields

$$H^{-1}\mathbf{g} = \sum_{i=1}^{N} \frac{\mathbf{g}^T \mathbf{v}_i}{\sigma_i} \mathbf{v}_i = \sum_{i=1}^{n} \frac{\sigma_i f_i + \eta_i}{\sigma_i} \mathbf{v}_i. \tag{6}$$

For the small indices $i \in \mathcal{I}_s$ there is a large number c with $\sigma_i > c\varepsilon$, and

$$\frac{\sigma_i f_i + \eta_i}{\sigma_i} = f_i + \frac{\eta_i}{\sigma_i} \approx f_i \pm \frac{1}{c}, \quad i \in \mathcal{I}_s,$$

where $1/c$ is small; these spectral components are indeed well reconstructed. For the larger indices $i \in \mathcal{I}_n$ we have instead $\sigma_i \ll \varepsilon$, and

$$\frac{\sigma_i f_i + \eta_i}{\sigma_i} = f_i + \frac{1}{\sigma_i} \eta_i, \quad i \in \mathcal{I}_n,$$

so that the noise contribution η_i is highly magnified.
We conclude that $H^{-1}\mathbf{g}$ is dominated by noise, and regularization has to deal with that, *e.g.*, by incorporating filter factors $\varphi_i \geq 0$ into the expansion (6):

$$\mathbf{f}_{\{\varphi_i\}} = \sum_{i=1}^{N} \varphi_i \frac{\sigma_i f_i + \eta_i}{\sigma_i} \mathbf{v}_i. \tag{7}$$

The filter factors φ_i should satisfy the following requirement: they should be close to 1 for $i \in \mathcal{I}_s$ – where stable reconstruction of f_i is possible – and they should be close to 0 for the critical $i \in \mathcal{I}_n$. It remains to discuss the impact of $i \in \mathcal{I}_t$. The corresponding φ_i should lie between 0 and 1, and the essential task of any regularizing algorithm is to choose these filter factors φ_i, $i \in \mathcal{I}_t$, in such a way that the approximation error

$$||\mathbf{f} - \mathbf{f}_{\{\varphi_i\}}||$$

is minimized. Note that optimal choice of these filter factors *a priori* is difficult (if not impossible) as it depends on the actual ratios f_i/η_i, $i \in \mathcal{I}_t$, which are unknown of course. Good *a posteriori* choices are necessary for a regularizing algorithm to be effective.

We emphasize that in image restoration applications, a similar notion of signal space, noise space and transition third space may be introduced in the Fourier domain: the blurred image $\bar{\mathbf{g}}$ is often close to being bandlimited, *i.e.*, its Fourier coefficients are small outside a symmetric interval $\mathcal{F}_s = \{\omega : 0 \leq |\omega| \leq \omega_s\}$. The expected Fourier coefficients of the white noise $\boldsymbol{\eta}$, however, have the same magnitude for all frequencies, ε say, and they will dominate the Fourier coefficients of $\mathbf{g} = \bar{\mathbf{g}} + \boldsymbol{\eta}$ in $\mathcal{F}_n = \{\omega : |\omega| \geq \omega_n > \omega_s\}$. Finally, $\mathcal{F}_t = \{\omega : \omega_s < |\omega| < \omega_n\}$ goes between $\mathcal{F}_s$ and $\mathcal{F}_n$.

4 Regularizing Properties of PCGNR

We continue our heuristic investigation of regularization in the special case of (P)CGNR, see also [10, 17]. More rigorous but less illustrative analyses can be found in Nemirovskii [14] and Plato [15].

We first consider the case $C = I$ (CGNR). Observe that the kth iterate $\mathbf{f}^{(k)}$ of Algorithm 1 assumes indeed a filtered SVD–expansion of form (7), *i.e.*,

$$\mathbf{f}^{(k)} = \sum_{i=1}^{N} \varphi_i^{(k)} \frac{\sigma_i f_i + \eta_i}{\sigma_i} \mathbf{v}_i,$$

where the filter factors $\varphi_i^{(k)}$ are given by the so–called *Ritz polynomials* r_k via

$$\varphi_i^{(k)} = 1 - r_k(\sigma_i^2). \tag{8}$$

The Ritz polynomials in turn may alternatively be introduced as a certain sequence of orthogonal polynomials (*cf.* [17]), or as the unique sequence of polynomials minimizing the quadratic form

$$F[p_k] = \sum_{i=1}^{N} p_k^2(\sigma_i^2)\ (\sigma_i f_i + \eta_i)^2 \tag{9}$$

over all polynomials p_k of degree k, normalized so that

$$p_k(0) = 1, \qquad k = 0, 1, 2, \ldots$$

We find the second characterization (9) more convenient. Rewriting the minimizer $F(r_k)$ in terms of the filter factors (8) yields

$$F[r_k] = \min_{\substack{\deg p_k = k, \\ p_k(0)=1}} F[p_k] = \sum_{i=1}^{N} (1 - \varphi_i^{(k)})^2 (\sigma_i f_i + \eta_i)^2. \tag{10}$$

Consider now the "weights" of the quadratic form F,

$$w_i = (\sigma_i f_i + \eta_i)^2.$$

According to the discrete Picard condition for $\bar{\mathbf{g}}$ and the white noise property of $\boldsymbol{\eta}$, we observe that the weights w_i will be heavier for $i \in \mathcal{I}_s$, that is, for spectral components corresponding to the signal subspace. The corresponding filter factors will therefore be the first to approach 1 as k increases.

On the other hand, $r_k(0) = 1$, and the weights w_i with $i \in \mathcal{I}_n$ are just of the order of ε. Thus, $r_k(\lambda)$ will remain close to 1 in an entire neighborhood of $\lambda = 0$ for moderate values of k, and the filter factors $\varphi_i^{(k)}$ for $i \in \mathcal{I}_n$ will therefore be close to 0 in the beginning of the iteration process.

Consequently, the $\varphi_i^{(k)}$ fulfill the filtering requirements that we have imposed for (7) – at least, up to some iteration number k'; at that point of the iteration CGNR has sufficient degrees of freedom when choosing the Ritz polynomials to cancel the contributions of $i \in \mathcal{I}_s$ to $F[r_k]$ in (10), and the contributions of $i \in \mathcal{I}_n$ start getting important in the minimization process (10) (a good way to think of these effects is in terms of an interpolation process; see for instance the analysis of Louis [13]). At this stage of the iteration, $r_k(\sigma_i^2)$, $i \in \mathcal{I}_n$ fixed, will significantly decrease and CGNR starts to reconstruct noise; the filtering requirements are violated and CGNR should ultimatively be stopped.

Again, we emphasize the role of $\mathcal{U}_t$: CGNR should not be stopped before the entire information from the signal subspace is reconstructed, but it must be stopped before it starts reconstructing the noise space components. When to stop exactly depends on the amount of signal information relative to noise that is contained in the transition subspace $\mathcal{U}_t$.

We now turn to the question of *preconditioning.* The standard idea of preconditioning is to speed up the convergence process; in the present context this should rather mean reducing the number of iterations required to reconstruct the information from the signal subspace. Depending on the robustness of our stopping rule, we may or may not wish to accelerate the convergence process as it reconstructs information from the transition subspace.

The number of iterations required to reconstruct the component of $\mathbf{f}$ in $\mathcal{U}_s$ mainly depends on the spread and the clustering of the singular values σ_i with $i \in \mathcal{I}_s$; the more clustered the singular values, the faster the convergence. As mentioned before, discrete convolution equations usually have densely distributed wide-spread spectra with little clustering, hence CGNR will require many iterations before the stopping rule is satisfied.

The preconditioner C should therefore cluster the spectrum of the mapping HC^{-1} restricted to the signal subspace U_s. On the other hand, the entire spectrum of HC^{-1} must not be clustered because this would combine the signal subspace and noise subspace; in such a case all iterates, right from the beginning, would be contaminated by noise, and the regularizing property of CGNR would be lost. Rather, HC^{-1} should act like H on the noise subspace, or, in other words, C should act like the identity on $\mathcal{U}_n$.

The remaining question about the behavior of C on the transition subspace $\mathcal{U}_t$ is delicate. It seems to us as if $C \approx I$ on $\mathcal{U}_t$ is the most appropriate for our purposes and applications: Due to the non-clustered part $\{\sigma_i : i \in \mathcal{I}_t\}$ of the spectrum of H, PCGNR will reach and pass the optimal stopping point quite slowly so that termination rules will work robustly. Of course, however, if the signal subspace $\mathcal{U}_s$ has been underestimated, then potential speedup will be lost with such a choice of C; we refer to the experimental section for examples.

We emphasize that the choice of C should depend on the actual noise level, because the definitions of $\mathcal{U}_s$, $\mathcal{U}_t$ and $\mathcal{U}_n$ depend on the noise variance ε^2. For smaller noise levels we can allow "more preconditioning" in the sense of "more clustering"; for large noise levels we can tolerate "little preconditioning" only.

The standard Level–2 preconditioner of Section 2 would correspond to "full preconditioning", giving as much clustering as possible with C. According to what we have said before, this choice of preconditioner is totally out of order for ill–posed problems because this would mix up the noise and signal subspaces.

In the next section we will show how the Level–2 preconditioner can be adapted to manipulate the level of preconditioning that we actually want. We also prove a result which essentially states that the singular values of the preconditioned matrix corresponding to the signal subspace cluster at 1, while the singular values corresponding to the noise subspace still cluster at 0.

5 Signal Subspace and the Truncation Parameter

As has been discussed in the previous section, a critical point of our preconditioning approach is the specification of the – so far only vaguely defined – signal subspace $\mathcal{U}_s$. Roughly speaking, the signal subspace should contain a major part of that right-hand side's fraction which comes from $\bar{\mathbf{g}}$, rather than from $\boldsymbol{\eta}$. Once $\mathcal{U}_s$ is specified, we use it to choose a *truncation parameter* τ. This parameter τ is chosen so that the eigenvalues of C satisfying $|\lambda_i| \geq \tau$ correspond to $\mathcal{U}_s$, with the remaining eigenvalues corresponding to $\mathcal{U}_n$ and $\mathcal{U}_t$. This truncation parameter is then used to construct our modified preconditioner, which we denote as C_τ, from C. We will postpone the specific description of C_τ until the next section, while concentrating here on how the signal subspace $\mathcal{U}_s$ is specified and how, from it, the truncation parameter τ is chosen.

As the discussion in Section 3 implies, a convenient subspace might be defined with Fourier vectors. At this point we recall the final remark of Section 3: it suggests computing the discrete Fourier transform of the right-hand side vector $\mathbf{g}$; its Fourier coefficients will decay to about the level of ε as the frequency increases (cf. Figures 3 and 5 for an example).

We suggest choosing ω_s as that particular frequency for which all Fourier coefficients relative to frequencies $|\omega| > \omega_s$ are less than ε in absolute value. As in Section 3, the signal space $\mathcal{U}_s$ is taken to be that space which is spanned by the Fourier vectors with frequencies of at most ω_s.

To compute our truncation parameter, τ, we thus compute the discrete Fourier transform of the right hand side $\mathbf{g}$, and find the point ε at which these coefficients level off. The index where this stagnation first begins indicates where the random errors start to dominate the right hand side. That is, we obtain a specification of the signal subspace. Thus, if p_τ is the index where the Fourier coefficients begin to level off, we then take the magnitude of the p_τ^{th} eigenvalue of C, $|\lambda_{p_\tau}|$, as the truncation parameter τ.

6 Modified Level–2 Preconditioner

In this section we describe how the Level–2 preconditioner can be modified to incorporate the amount of preconditioning we want; and, in addition, we provide some theoretical results on the effectiveness of the preconditioner. It was suggested in Section 4 how $C \equiv \mathcal{C}_2(H)$ is to be modified, but it is useful to state an algorithm which explicitly shows how the modified Level–2 preconditioner is to be constructed.

First, to obtain an explicit description of our preconditioner, recall that C is a BCCB matrix, and hence can be easily diagonalized using the 2-D FFT. In fact, when using C as a preconditioner, one needs only its eigenvalues (see Algorithm 2.) Now recall further from Section 4, that for ill-posed problems we want the modified preconditioner to act like C on $\mathcal{U}_s$, and like the identity on $\mathcal{U}_n$ and on $\mathcal{U}_t$.

This is equivalent to locating the eigenvalues of C which correspond to $\mathcal{U}_n$ and to $\mathcal{U}_t$ and replacing them with ones.

As discussed in Section 5, we choose a truncation parameter τ such that eigenvalues of C satisfying $|\lambda_i| \geq \tau$ correspond to $\mathcal{U}_s$, with the remaining eigenvalues corresponding to $\mathcal{U}_n$ and $\mathcal{U}_t$. To simplify our discussion, we assume that the eigenvalues of C are λ_i with $|\lambda_1| \geq |\lambda_2| \geq \cdots \geq |\lambda_{mn}|$, and we define the *truncation index* p_τ to be the largest integer p such that $|\lambda_p| \geq \tau$. Moreover, since our preconditioner depends on the parameter τ, we denote it by C_τ.

Now, as the above remarks suggest, given the tolerance τ, to construct C_τ we first find the eigenvalues λ_i of C. Then C_τ is simply the BCCB matrix with eigenvalues

$$\tilde{\lambda}_i = \begin{cases} \lambda_i & \text{if } |\lambda_i| \geq \tau \\ 1 & \text{if } |\lambda_i| < \tau \end{cases} .$$

This discussion is summarized in the following algorithm.

Algorithm 3. Construction of the modified **Level–2** preconditioner for 2-D problems. *Let H be the $BTTB(m,n)$ matrix given as in* (2), *and let τ be chosen by the method described in Section* 5. *This algorithm computes the eigenvalues of the BCCB preconditioner C_τ, using the 2-D FFT, and stores them in the diagonal matrix Λ.*

Use Algorithm 2 to compute the eigenvalues λ_i of C.

If $|\lambda_i| < \tau$, replace λ_i with 1.

The cost of finding C_τ, and for solving systems with C_τ, is on the same order as that for C (see remarks following Algorithm 2.)

We now turn to the convergence analysis of the PCGNR method with C_τ used as a preconditioner. As mentioned earlier, the convergence rate of the PCGNR depends on the distribution of the singular values of HC_τ^{-1}. In particular, if the singular values of HC_τ^{-1} corresponding to the signal subspace $\mathcal{U}_s$ cluster around one, then the convergence will be rapid.

Here, as in [4], we assume that the entries $h_{i-j,k-l}$ of H are obtained from a doubly indexed sequence $\{h_{\mu,\nu}\}$, and that

$$\sum_{\mu=-\infty}^{\infty} \sum_{\nu=-\infty}^{\infty} |h_{\mu,\nu}| \leq K < \infty. \tag{11}$$

Thus we immediately have

$$||H||_2^2 \leq ||H||_1 ||H||_\infty \leq K^2, \tag{12}$$

where, as usual, $||H||_1$ and $||H||_\infty$ denote the column and row sum norms of H, respectively.

The theoretical analysis to follow concerns the limiting process where m and n tend to infinity, while the noise variance ε^2 is assumed to be fixed. Such an analysis reflects the process of increasing the level of discretization. Note that essentially, our choice of τ only depends on ε but not on m or n.

The following lemma relating C and H is proved in [4].

Lemma 6.1. *Let H satisfy* (11). *Then for all $\delta > 0$, there exist $M, N > 0$, such that for all $m > M$ and $n > N$,*

$$H - C = U + V$$

where

$$\text{rank}(U) = O(m) + O(n)$$

and

$$||V||_2 < \delta.$$

We now consider the modified preconditioner C_τ. The following lemma establishes how well C_τ approximates H.

Lemma 6.2. *Let H satisfy* (11), *assume $\tau > 0$, and let C_τ be constructed as above. Then for all $\delta > 0$, there exist $M, N > 0$, such that for all $m > M$ and $n > N$,*

$$H - C_\tau = U + V + W - P_\tau$$

where P_τ is the orthogonal projector onto the direct sum $\mathcal{U}_n \oplus \mathcal{U}_t$,

$$\text{rank}(U) = O(m) + O(n),$$

$$||V||_2 < \delta,$$

and W is a matrix satisfying $W = P_\tau W P_\tau$, and

$$||W||_2 < \tau.$$

Proof. For the spectral decomposition of BCCB matrices we make use of the 2–D Fourier transform matrix $\mathcal{F}$ [12], and we assume without loss of generality that the columns of $\mathcal{F}^*$ contain the eigenvectors of C in such an ordering that the corresponding eigenvalues of C are in non-increasing order. In particular, if p_τ is the truncation index corresponding to τ, the final $mn - p_\tau$ columns of $\mathcal{F}^*$ contain an orthonormal basis for $\mathcal{U}_n \oplus \mathcal{U}_t$.

Observe from Lemma 6.1 that we can write

$$H - C_\tau = H - C + C - C_\tau = U + V + W - P_\tau$$

where U and V are from Lemma 6.1, and

$$W = \mathcal{F}^* \text{diag}(0, \ldots, 0, \lambda_{p_\tau+1}, \ldots, \lambda_{mn})\mathcal{F},$$

$$P_\tau = \mathcal{F}^* \text{diag}(0, \ldots, 0, 1 \ldots, 1)\mathcal{F}.$$

From this it follows that P_τ is the orthogonal projector onto $\mathcal{U}_n \oplus \mathcal{U}_\tau$, and obviously $W = P_\tau W P_\tau$. Furthermore, as $\mathcal{F}$ is a unitary matrix,

$$\|W\|_2 \leq \lambda_{p_\tau+1} < \tau.$$

The properties of U and V now follow from Lemma 6.1. □

We now present our clustering result for the singular values of the preconditioned matrix HC_τ^{-1}.

Theorem 6.1. *Let H satisfy (11), assume $\tau > 0$ and let C_τ defined as above. Then given $\delta > 0$, there exist $M, N > 0$, such that for all $m > M$ and $n > N$,*

$$HC_\tau^{-1} - (I - P_\tau) = U + V + W$$

where U and V satisfy

$$\text{rank}(U) = O(m) + O(n)$$

and

$$||V||_2 < \delta, \qquad ||W||_2 < \tau.$$

Thus at most $O(m) + O(n)$ of the singular values of the preconditioned matrix

$$HC_\tau^{-1}$$

lie outside the two intervals $(1 - \delta - \tau, 1 + \delta + \tau) \cup [0, \delta + \tau)$.

Proof. First, notice that from Lemma 6.2 we can write

$$HC_\tau^{-1} - (I - P_\tau) = (H - C_\tau)C_\tau^{-1} + P_\tau = (\tilde{U} + \tilde{V} + \tilde{W})C_\tau^{-1} + P_\tau(I - C_\tau^{-1}),$$

where $\tilde{U}$, $\tilde{V}$ and $\tilde{W}$ are given by Lemma 6.2. By definition, $P_\tau(I - C_\tau^{-1}) = 0$, so that we obtain the identity

$$HC_\tau^{-1} - (I - P_\tau) = U + V + W$$

where we have let $U = \tilde{U}C_\tau^{-1}$, $V = \tilde{V}C_\tau^{-1}$ and $W = \tilde{W}C_\tau^{-1}$. Note that the particular form of $\tilde{W}$ derived in Lemma 6.2 implies that $W = \tilde{W}$ and hence W satisfies the given assertion. Furthermore,

$$\text{rank}(U) = \text{rank}(\tilde{U}) \leq O(m) + O(n).$$

Now observe that, from the definition of C_τ, its minimum eigenvalue is bounded by $\min\{1, \tau\}$. Thus, assuming $\tau \leq 1$ and using Lemma 6.2, we obtain

$$\|V\|_2 \leq ||\tilde{V}||_2||C_\tau^{-1}||_2 \leq \tilde{\delta}/\tau.$$

Now since the $\tilde{\delta}$ can be chosen arbitrarily small, it follows that $\|V\|_2 < \delta$.

Then, as $I - P_\tau$ is an orthogonal projector with all its singular values being either 0 or 1, it follows from [19] that at most $O(m) + O(n)$ singular values of HC_τ^{-1} lie outside the union of the intervals $(1 - \delta - \tau, 1 + \delta + \tau) \cup [0, \delta + \tau)$. □

We note that using a standard error analysis of the conjugate gradient method, it can be shown (see, *e.g.*, [4]) that our method here will give the best regularized approximation to **f** in at most $O(m) + O(n)$ steps for m and n sufficiently large. Also it follows from Theorem 6.1 that $HC_\tau^{-1} \approx I$ on $\mathcal{U}_s$ whereas $HC_\tau^{-1} \approx H \approx 0$ on $\mathcal{U}_n$; in other words, C_τ is a good preconditioner on the signal subspace and acts like the identity on the noise subspace, as desired. We also emphasize that for moderately ill-posed problems with little noise where we can choose $\tau \approx 0$, we can make use of "full preconditioning" and then we obtain convergence in at most $O(m) + O(n)$ steps.

7 Numerical Results for 2-D Problems

In this section we provide some numerical experiments illustrating the effectiveness of our methods. The examples we use are model problems in image restoration computations [12], and are of the form

$$\mathbf{g} = H\mathbf{f} + \boldsymbol{\eta},$$

where H is a BTTB matrix representing the blurring in the image. All computations were performed on the Cray Y-MP at the North Carolina Supercomputing Center.

We begin by generating a known 64×64 image shown in Figure 1, and form the vector $\mathbf{f}$ of dimension 4096 by row ordering (see [12]) the original image. We then choose the ill-conditioned 4096×4096 blurring matrix H as

$$h_{i-j,k-l} = \begin{cases} \exp\{-0.1((i-j)^2 + (k-l)^2)\} & -5 \le i-j, k-l \le 5 \\ 0 & \text{otherwise} \end{cases}, \tag{13}$$

and construct the blurred, noisy image by forming the vector $\mathbf{g} = H\mathbf{f} + \boldsymbol{\eta}$ (see Figure 2). The entries of the vector $\boldsymbol{\eta}$ are generated randomly from a normal distribution with a mean of 0.0 and a variance of $\varepsilon^2 = 1.0$, and scaled to obtain specific noise levels. Two noise levels are considered, $\|\boldsymbol{\eta}\|_2/\|H\mathbf{f}\|_2 = 10^{-3}$ and 10^{-2}, *i.e.* 0.1% and 1% noise.

To choose the truncation parameter τ, we use the scheme suggested in Section 5. Specifically, we compute the discrete Fourier transform of the right hand side $\mathbf{g}$, and find the level ε at which these coefficients level off. The index where this stagnation first begins indicates where the random errors start to dominate the right hand side. That is, we obtain a specification of the signal subspace. Thus, if i is the index where the Fourier coefficients begin to level off, we then take the magnitude of the pth eigenvalue of C, $|\lambda_p|$, as the truncation parameter τ used in Section 6 for constructing the preconditioner C_τ.

Example 1: 0.1% noise.

In this first example, we consider 0.1% noise. Figure 3 shows a plot of the eigenvalues of C along with a plot of the Fourier coefficients of the right hand side. Here we see that the Fourier coefficients begin to stagnate when $p \approx 1450$. Thus, we take $|\lambda_{1450}| \approx 0.0032$ as the truncation parameter τ.

In Figure 4, we show the relative errors for several iterations of PCGNR with our preconditioner and with no preconditioner. The minimal relative error for PCGNR, with our preconditioner, is attained at $k = 30$ iterations, with a relative error value of ≈ 0.1698. On the other hand, when no preconditioner is used, then the minimal relative error occurs at $k = 189$ iterations, with a relative error value of ≈ 0.1654. These results are summarized in Table 1. Note from Figure 4 that the convergence rate of PCGNR, with our preconditioner, is significantly faster than the CGNR with no preconditioning. Each method produces a similar minimal relative error.

	0.1% noise		1% noise	
	no prec.	prec.	no prec.	prec.
iterations	189	30	43	6
min. error	0.1654	0.1698	0.2236	0.2276

Table 1: Summary of numerical results from Examples 1 and 2.

Example 2: 1% noise.

In this example we generate 1% noise. Again, to choose the truncation parameter, we consider the plots of the eigenvalues of C and the corresponding Fourier coefficients of the right hand side, shown in Figure 5. From these plots, we see that the Fourier coefficients begin to level off when $p \approx 500$.

Using $|\lambda_{500}| \approx 0.0233$ as the truncation parameter τ, the relative errors for several iterations of PCGNR with our preconditioner, and with no preconditioner, are shown in Figure 6. In this case, the minimal relative error computed by PCGNR occurs at $k = 6$ iterations with a relative error value of ≈ 0.2276, whereas the minimum relative error for CGNR occurs at $k = 43$ iterations with a relative error value of ≈ 0.2236. These results are also summarized in Table 1.

Choosing an "optimal" truncation parameter is critical to the effectiveness of our preconditioners. We illustrate this in the following two examples, where we use the data from Example 2 with 1% noise.

Example 3: Tolerance too large.

In this example, we choose the truncation parameter to be $\tau = |\lambda_{250}| \approx 0.1539$. This has the effect of underestimating the signal subspace (see Figure 5), and, as remarked in Section 4, we expect the potential speedup of the preconditioner to be lost. This is easily seen in Figure 7, where we show the relative error for each iteration of the PCGNR with this preconditioner (dashed line) as well as with no preconditioner (solid line). In this case, the minimal relative error is reached at $k = 41$ iterations, with a value of 0.2238. This should be compared to Figure 6 where the optimal truncation parameter was used.

Example 4: Tolerance too small.

In this example we take the truncation parameter to be $|\lambda_{1000}| \approx 0.0077$. In this case we have, effectively, overestimated the signal subspace (see Figure 5). Thus we would expect the noise to dominate, and hence we will not be able to obtain as accurate a solution as in Example 2. This can also be seen in Figure 7, where the dash-dot line is a plot of the relative errors for each iteration of the PCGNR with the preconditioner of this example. Here we reach the minimum relative error at iteration 9, with a large relative error value of 0.3005. Clearly, the optimal truncation parameter in Example 2 produces a much more accurate solution with fewer iterations.

The restored image for Example 1 (0.1% noise) using our algorithm is given in Figure 8. Our examples illustrate advantages, as well as some problems, with choosing a good truncation parameter for our preconditioner. We see that if the tolerance is chosen well, then we get a significant speed-up in convergence to a minimal relative error solution. But the above examples also indicate, as was eluded to in Section 4, that the relative error of the approximate solution obtained from the PCGNR iterations can be very sensitive to the choice of the truncation parameter due to the ill-conditioned nature of C_τ. In particular, if the truncation parameter τ is underestimated then we may not be able to compute an accurate approximation to the exact solution. This indicates that one should choose $C \approx I$ on $\mathcal{U}_t$, as was suggested in Section 6. Moreover, if it is not clear where the signal subspace ends, then it appears to be best to overestimate the truncation tolerance τ, rather than underestimate it.

Finally, we remark that we have not addressed the very important, and difficult, problem of deciding when to stop the PCGNR iterations. There has recently been much research devoted to this topic, and we refer the interested reader to the discussion in [10]. We emphasize that if one has a good stopping criterion for the PCGNR, then our approach to preconditioning may offer significant improvement on convergence rate for ill-posed problems of the type discussed in this paper.

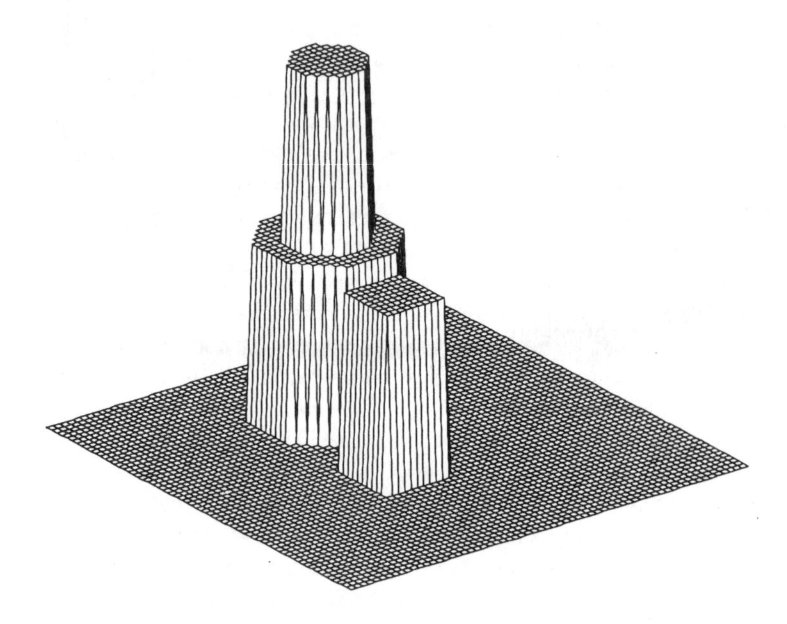

Figure 1: Original image.

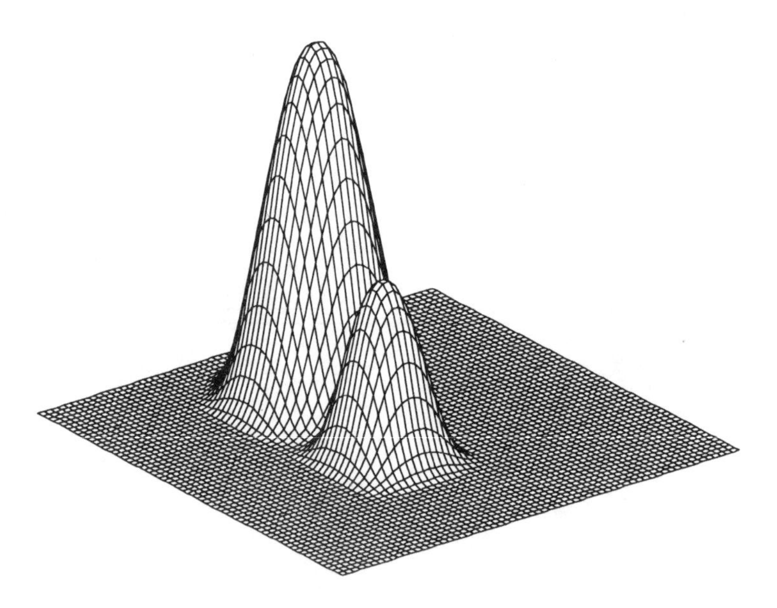

Figure 2: Blurred image with 0.1% noise.

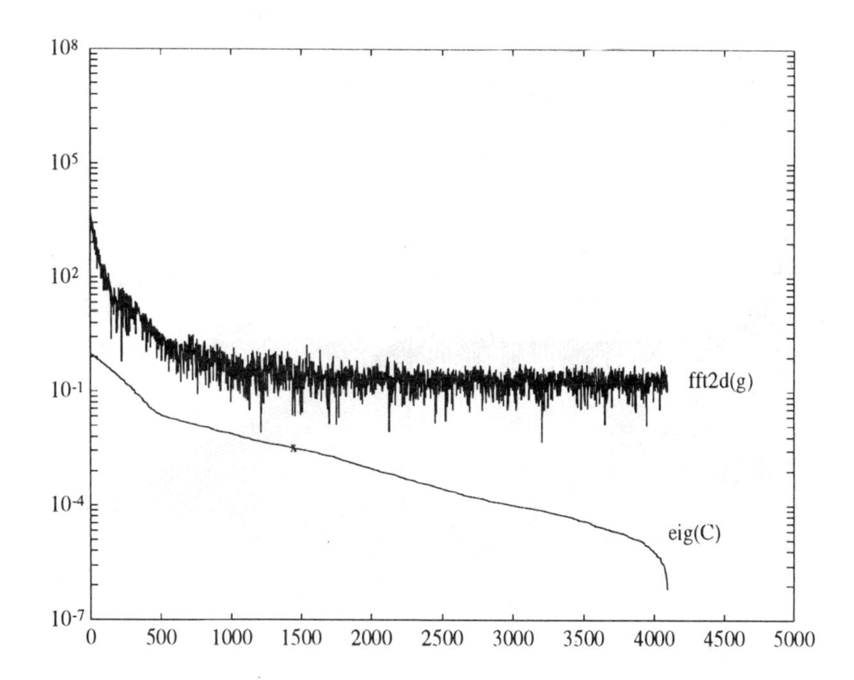

Figure 3: Eigenvalues of C and Fourier transform of $\mathbf{g}$ for 0.1% noise.

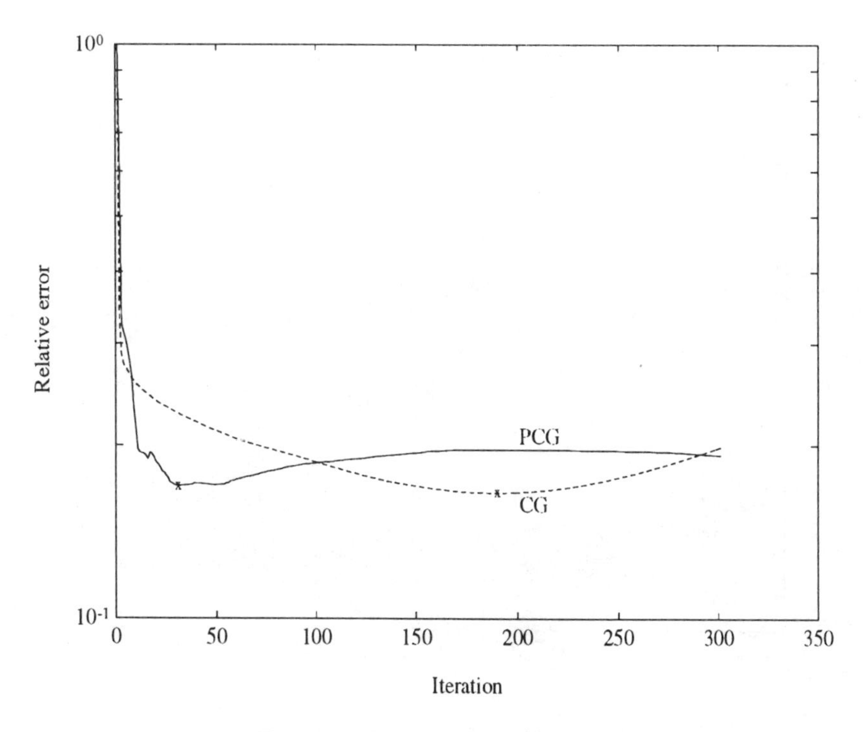

Figure 4: Errors for 0.1% noise.

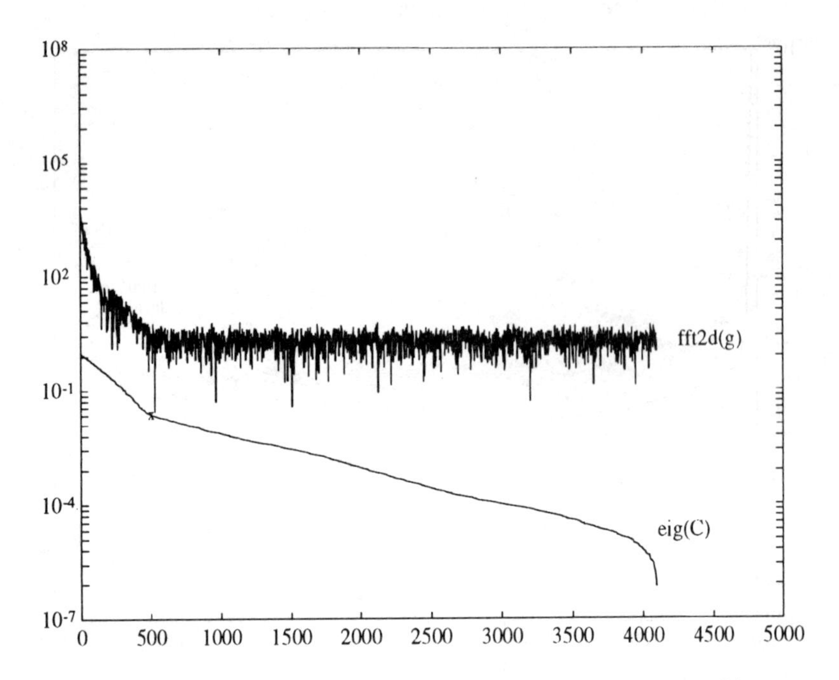

Figure 5: Eigenvalues of C and Fourier transform of $\mathbf{g}$ for 1% noise.

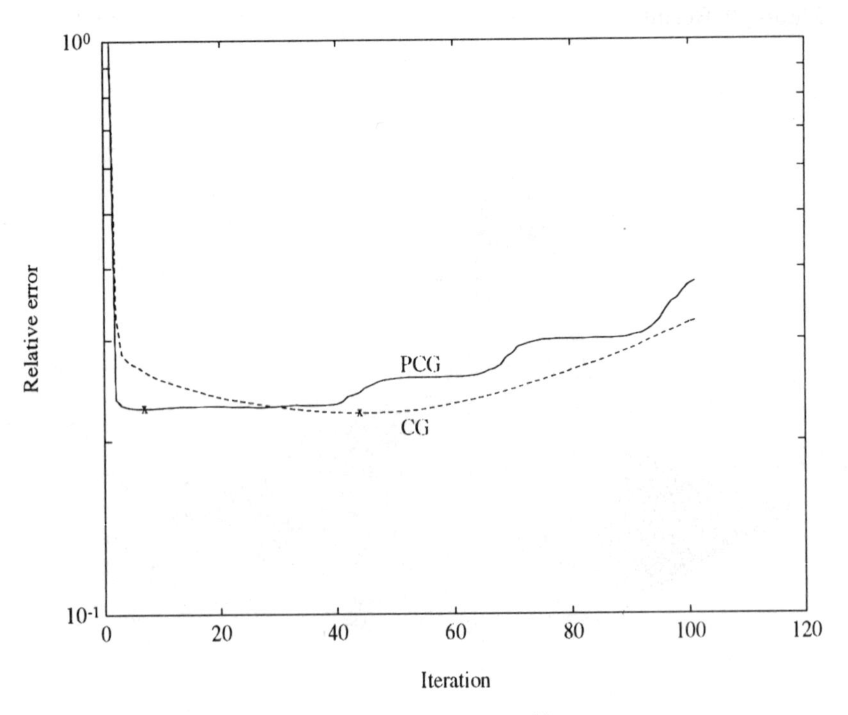

Figure 6: Errors for 1% noise.

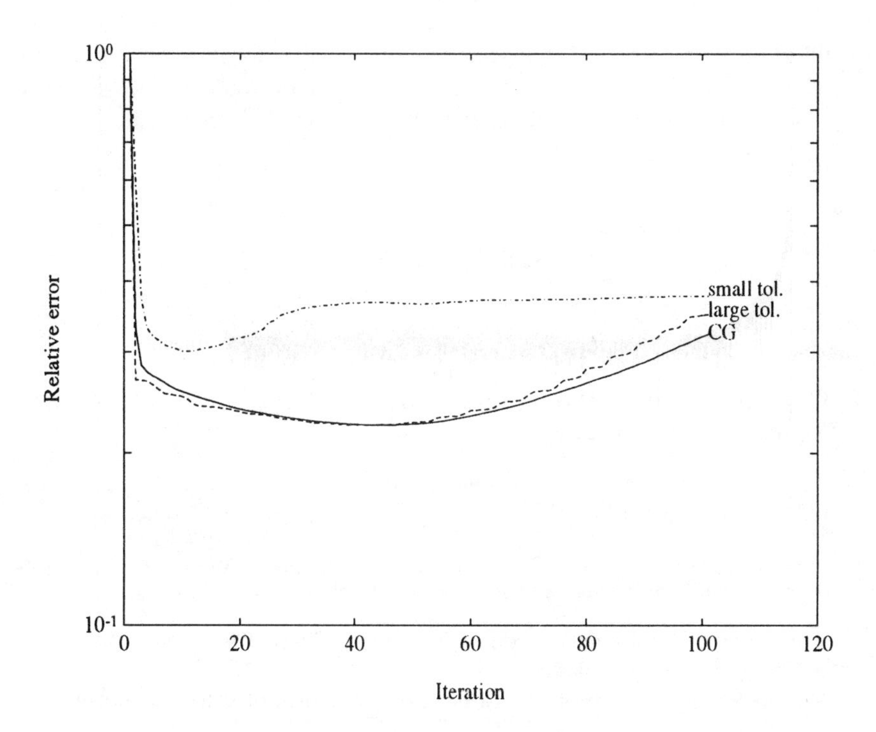

Figure 7: Results when tolerance τ is chosen too large or too small.

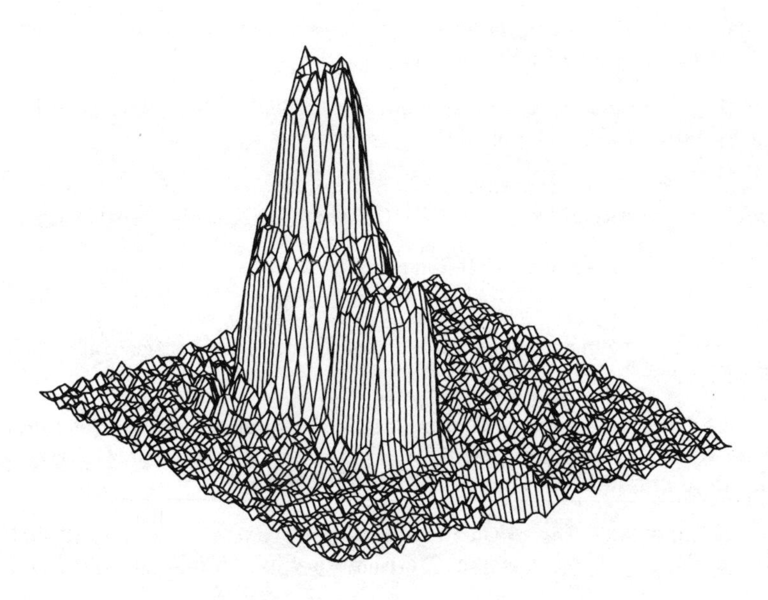

Figure 8: Restored image in Example 1.

Acknowledgement. Much of this work was done while the authors were participating in the *Year on Applied Linear Algebra* at the Institute for Mathematics and its Applications, University of Minnesota, Minneapolis, MN 55455.

References

[1] A. BJÖRCK, *Least Squares Methods*, in Handbook of Numerical Methods, ed. P. Ciarlet and J. Lions, Elsevier/North Holland Vol. 1, 1989.

[2] R. CHAN AND X. JIN, *A Family of Block Preconditioners for Block Systems*, SIAM J. Sci. Stat. Comp., to appear.

[3] R. CHAN, J. NAGY AND R. PLEMMONS, *Circulant preconditioned Toeplitz least squares iterations*, preprint (1991), SIAM J. Matrix Anal. Appl., to appear.

[4] R. CHAN, J. NAGY AND R. PLEMMONS, *FFT-based preconditioners for Toeplitz-block least squares problems*, preprint (1992), SIAM J. Numer. Anal., to appear.

[5] T. CHAN, *An optimal circulant preconditioner for Toeplitz systems*, SIAM J. Sci. Stat. Comp., V9 (1988), pp. 766–771.

[6] T. CHAN AND J. OLKIN, *Preconditioners for Toeplitz-block matrices*, preprint (1992).

[7] G. GOLUB AND C. VAN LOAN, *Matrix Computations*, Johns Hopkins Univ. Press, Baltimore, MD, 2nd Ed., 1989.

[8] C. GROETSCH, *The Theory of Tikhonov Regularization for Fredholm Equations of the First Kind*, Pitman Publishing, Boston, 1984.

[9] M. HANKE, *Accelerated Landweber iterations for the solution of ill-posed problems*, Numer. Math., V60 (1991), pp. 341–373.

[10] M. HANKE AND P. C. HANSEN, *Regularization Methods for Large-Scale Problems.* Technical University of Denmark, UNI•C Report UNIC-92-04, August 1992.

[11] P. C. HANSEN, *Truncated SVD solutions to discrete ill-posed problems with ill-determined numerical rank*, SIAM J. Sci. Stat. Comp., V11 (1990), pp. 503-518.

[12] A. K. JAIN, *Fundamentals of Digital Image Processing*, Prentice-Hall, Englewood Cliffs, NJ 1989.

[13] A. K. LOUIS, *Convergence of the conjugate gradient method for compact operators*, in inverse and ill-posed problems, H. W. Engl and C. W. Groetsch (Eds.), Academic Press, Boston, New York, London, 1978, pp. 177–183.

[14] A. S. NEMIROVSKII, *The regularizing properties of the adjoint gradient method in ill-posed problems*, USSR Comput. Math. Math. Phys., V26,2 (1986), pp. 7–16.

[15] R. Plato, *Optimal algorithms for linear ill-posed problems yield regularization methods*, Numer. Funct. Anal. Optim., V11 (1990), pp. 111–118.

[16] G. Strang, *A proposal for Toeplitz matrix calculations*, Stud. Appl. Math., V74 (1986), pp. 171–176.

[17] A. van der Sluis and H. van der Vorst, *SIRT- and CG-type methods for the iterative solution of least squares problems*, Linear Alg. Appl., V130 (1990), pp. 257–302.

[18] A. van der Sluis and H. van der Vorst, *The rate of convergence of conjugate gradients*, Numer. Math., V48 (1986), pp. 543–560.

[19] J. Wilkinson, *The Algebraic Eigenvalue Problem*, Clarendon Press, Oxford, 1965.

Block-ADI preconditioners for solving sparse nonsymmetric linear systems of equations

Sangback Ma *† *and Youcef Saad* ‡§

Abstract. There is currently a renewal of interest in the Alternating Direction Implicit (ADI) algorithms as preconditioners for iterative methods for solving large sparse linear systems, because of their suitability for parallel computing. However, the classical ADI iteration is not directly applicable to finite element (FE) matrices, as it tends to converge too slowly for 3-D problems, and as the selection of adequate acceleration parameters, remains a difficult task. In this paper we propose a Block-ADI approach, which overcomes some of these problems. In particular we derive a simple inexpensive heuristic for selecting the acceleration parameters. The new approach can be viewed as a combination of the classical ADI method and a domain decomposition approach.

Key words. preconditioning, ADI, block ADI, preconditioned GMRES, domain decomposition.

AMS (MOS) subject classification: 65F10.

1 Introduction

Iterative solutions of large sparse linear systems by Krylov subspace methods, require the use of preconditioning techniques in order to achieve convergence in a reasonable number of iterations. When implemented on parallel computers, standard preconditioners based on incomplete factorization, such as ILU [14] or the more elaborate versions such as ILUT[17, 24], using a wavefront or level-scheduling approach realize a reasonable speed-up on vector computers or parallel computers

*This research was supported by NIST grant 60NANB2D11272 and by The Minnesota Supercomputer Institute.

†University of Minnesota, Computer Science Department, 4-192 EE/CSci Building, 200 Union Street S.E., Minneapolis, MN 55455.

‡This research was supported by NIST grant 60NANB2D11272 and by The Minnesota Supercomputer Institute.

§University of Minnesota, Computer Science Department, 4-192 EE/CSci Building, 200 Union Street S.E., Minneapolis, MN 55455.

with a small number of processors [1, 2, 5]. However, the lengths of the wavefronts are not uniform, and this contributes to poor load balancing, and limits the maximum achievable speed-up due to Amdahl's law. Alternatively, parallelism can also be obtained through *multicoloring.* For example, if the matrix has property A, as is the case for the standard 5-point matrices obtained from centered finite difference (FD) discretizations of elliptic partial differential equations (PDE's), there is a partition of the grid-points in two disjoint subsets such that the unknowns of any one subset are only related to unknowns of the other subset. This one enables to produce a reordered matrix having a block-tridiagonal structure, where the diagonal blocks are diagonal matrices. There are several different ways of exploiting this structure. For example, the unknowns associated with one of the subsets can be easily eliminated, and the resulting reduced system is often well-conditioned. This 'two-coloring' often referred to as a red-black or checkerboard ordering, can be generalized to arbitrary sparse matrices by using multi-coloring, and the generalizations of the standard ILU techniques based on such ideas can be easily derived [19]. Polynomial preconditioners are quite appealing because of the requirement that only matrix – vector product operations need to be optimized; a task that is reasonably straightforward. They can also always be combined with other preconditioners to improve their overall performance, and this may constitute their best possible use, at least in the non-Hermitian case.

Another standard approach that has been employed in the past is domain decomposition. Most domain decomposition preconditioners rely on an efficient solution technique for the Schur complement. If we wish to achieve high speed-ups we must increase the number of subdomains, and as a result the Schur complement will become bigger and more complex. This eventually becomes difficult to program and, in addition, will yield diminishing returns.

Going back to standard preconditioners, we recall that historically SSOR was used first as a preconditioner to the conjugate gradient methods [3, 4] well before incomplete factorization techniques became popular. Similarly, standard relaxation or block relaxation techniques have often provided easy-to-implement and yet reasonably efficient preconditioners. However, insufficient work has been done on the parallel implementations of these techniques. In [18], SOR and SSOR and multicoloring techniques were combined and compared with 'good' ILUT implementations. The main result in [18] is that these types of techniques can be quite efficient, and far superior to the standard preconditioners on some problems, provided a multi-step approach is used, namely provided that $k > 1$ steps of SOR or SSOR are taken at each preconditioning operation instead of just one as is usually done, where k is some parameter.

In this paper we investigate an ADI type iteration viewed from a *domain decomposition* perspective. Our goal is to obtain a method whose computational structure is as amenable to parallel computing as the ADI iteration is. In addition we would like the method to be applicable to finite element applications, possibly for 3-D problems, just as in a domain decomposition approach.

2 Classical ADI Methods

The ADI method was introduced by Peaceman and Rachford [15] in 1955, to solve the discretized boundary value problems arising from elliptic and parabolic PDEs. The finite difference discretization of the model elliptic problem

$$\begin{aligned} -\Delta u &= f, \ \Omega = [0,1] \times [0,1] \\ u &= 0 \ on \ \delta\Omega \end{aligned} \tag{1}$$

with 5-point centered finite difference discretization, with $n+2$ mesh-points in the x direction and $m+2$ points in the y direction, leads to the solution of a linear system of equations of the form

$$Au = b \tag{2}$$

where A is a matrix of dimension $N = n \times m$. Without loss of generality and for the sake of simplicity, we will assume for the remainder of this paper that $m = n$, so that $N = n^2$.

Writing the discretization in x and y direction into matrices H and V respectively, leads to a linear system of equations

$$(H + V)u = b \tag{3}$$

where both H and V are sparse and possess a special structure. In particular, with suitable reordering, H and V are *tridiagonal.*

Starting with some initial guess u_0, the Alternative Direction Implicit procedure for solving (3) generates a sequence of approximations $u_i, i = 1, 2, \ldots$ given by the following algorithm

Algorithm **2.1. Peaceman Rachford(PR) ADI**

$$(H + \rho_i I)u_{i+1/2} = -(V - \rho_i I)u_i + b \tag{4}$$

$$(V + \rho_i I)u_{i+1} = -(H - \rho_i I)u_{i+1/2} + b, \tag{5}$$

where the ρ_i's are positive parameters. The ADI method, applied to positive definite systems, was extensively studied in the 1950s and 1960s, see, e.g, the books of Varga [21] and Wachspress [22]. In this case H and V have real eigenvalues and the following is a summary of the main ADI results in this situation.

1. Any stationary iteration($\rho_i = c > 0$, for all i) is convergent if H and V are symmetric positive definite.

2. For the model problem the asymptotic rate of convergence of ADI with optimal fixed ρ is equal to that of SSOR with optimal ω, so the spectral radius is $\omega_b - 1$. Since each ADI iteration takes more work than an SSOR iteration, a sequence of parameters ρ_i is often used in order to compete with SSOR ([21]).

3. The rate of convergence of ADI can be appreciably increased by the application of sequence of parameters, ρ_i, that are used in cyclic order. The theory of the convergence and the selection of good parameters when more than

one one cycle is used, has not been fully developed. For those cases when $HV = VH$, a satisfactory theory does exist [6]. For the model problem with a single optimal ρ the time complexity is $O(n^3)$. In [15] a procedure is described to reduce the time complexity to $O(n^2 \log n)$ with a sequence of ρ parameters.

For 3-D problems Douglas[7] proposed a variant of the classical ADI, which has better convergence behavior than the classical ADI. Let $A = H + V + W$, where H, V , and W contains the x-, y-, z-, directional derivatives, respectively.

ALGORITHM **2.2. Douglas Rachford(DR) ADI**

$$(H + \rho_i I)u_{i+1/3} = -(H + 2V + 2W - \rho_i I)u_i + 2b \tag{6}$$

$$(V + \rho_i I)u_{i+2/3} = -(H + V + 2W - \rho_i I)u_i - Hu_{i+1/3} + 2b \tag{7}$$

$$(W + \rho_i I)u_{i+1} = -(H + V + W - \rho_i I)u_i - Hu_{i+1/3} - Vu_{i+2/3} + 2b \tag{8}$$

The convergence behavior of this algorithm is different than in the two-dimensional case. For example for fixed $r > 0, \rho_i = r, \text{foralli}$, three dimensional mesh regions can be exhibited for which this variant of ADI fails. Also in practice the convergence is very slow, and as a result it has rarely been used. There are alternative formulations which sweep through planes as opposed to lines of the domain.

More recently, the focus has turned into the implementation and efficiency issues of ADI methods on parallel computers. The degree of parallelism of the algorithm is of the order of the number of the grid points, but the best complexity that can be achieved for each step is $O(\log n)$ using cyclic reduction in both directions to solve the tridiagonal systems [12, 10]. Other strategies exist which combine divide-and-conquer tridiagonal solvers and cyclic reduction.

3 Block versions of ADI

Since H and V depend on the finite difference discretizations of original PDEs, the classical ADI is not defined for FE matrices. For example, the piecewise linear shape functions on triangles give rise to 7-point matrices, for which there is no natural splitting of A into the sum of two matrices H and V that are both tridiagonal, or defined discretizations of one-dimensional operators. The question then arises as to how to generalize the classical ADI for finite elements applications. There are several options available. In this paper we will simply use a technique which is based on recasting the Peaceman Rachford ADI in the framework of *domain decomposition* methods.

3.1 The classical ADI and domain decomposition

In Algorithm 2.1 H is the discretization matrix of x-directional derivatives. In terms of domain decomposition the domain is decomposed into *horizontal* lines.

Then H is obtained by applying the original PDE on the subdomains, while imposing the Neumann boundary conditions on the *vertical* sides. Similarly for V. After H and V are found we could write A as

$$A = H + (A - H) = V + (A - V).$$

These two splittings of A are used in each of two stages of the iteration (4). A parameter ρ_i was added to the diagonals of H and V as an acceleration parameter. In other words ADI can be viewed as an extreme case of domain decomposition in the plane, where the subdomain consists of nonoverlapping horizontal rectangles consisting of one line each. We can also view ADI as a means of using a domain decomposition strategy to reduce two-dimensional domains into 1-dimensional subdomains. By alternating between the x and y directions we can achieve the overlapping between the domains that is desirable in domain decomposition. As we noted earlier in domain decomposition the convergence deteriorates if the number of subdomains increases and there is no overlap between the subdomains. By the *alternation* we hope to achieve the equivalent effect of overlapping subdomains.

3.2 A Block-ADI Algorithm

We have seen in the previous discussions that the two stages of the classical ADI are characterized by the way in which the matrix A is split in two additive components. It is natural to think of considering the subdomains of horizontal/vertical *stripes* consisting of a few, say k, lines, instead of just one line. The same procedure as in the classical ADI can then be defined. Let us call ADI(k) this variant of ADI, and let $H^{(k)}$ and $V^{(k)}$ denote the matrices obtained by applying the original PDE on this decomposition of the domains. In essence, for each of the two domain partitionings, these matrices are obtained from the original matrix by neglecting the interactions between grid points across interfaces, or rather replacing them with Neuman boundary conditions. Then A is split as

$$A = H^{(k)} + (A - H^{(k)}) = V^{(k)} + (A - V^{(k)})$$

from which we can define our block ADI procedure, denoted by ADI(k).

ALGORITHM **3.1. ADI(k)**

$$(H^{(k)} + \rho_i I)u_{i+1/2} = -(A - H^{(k)} - \rho_i I)u_i + b \tag{9}$$

$$(V^{(k)} + \rho_i I)u_{i+1} = -(A - V^{(k)} - \rho_i I)u_{i+1/2} + b \tag{10}$$

- Note that ADI(k) is defined for FE matrices as well as for FD matrices.
- In a parallel environment the simplest form of parallelism is obtained by processing the domains independently, which will naturally achieve a speed-up of n/k (in case $m = n$). Further speed-ups can also be achieved. Although the matrices $H^{(k)}$ and $V^{(k)}$ are not tridiagonal for $k > 1$, they are banded with a bandwidth of $2k + 1$ and parallel banded solvers [11] can be used for each subdomain independently to yield higher speed-ups.

- If we assume an exact LU factorization is used to solve (9), we expect that the cost will be of the order k^2 times that of the classical ADI. However, since $H^{(k)}$ and $V^{(k)}$ will be closer to A, or fewer interfaces are neglected in $H^{(k)}$ and $V^{(k)}$, the ADI(k) is likely to converge faster. As a result there is a tradeoff between the cost due to the number of ADI iterations and the cost of solving each of the equations (9) and (10). An optimal value of k should achieve a balance between these two costs.

- For 3-D problems we expect the 3-D equivalent of Block-ADI to perform better than DR–ADI for large k. The main reason why the DR–ADI converges so slowly is that H, V, and W are each too poor approximations to A. In terms of the splitting of A the spectral radius associated with the ADI iteration will be smaller as each of H, V or W gets closer to A. The matrices $H^{(k)}, V^{(k)}$, and $W^{(k)}$ are likely to be better approximations to A, for larger k. As in the 2-D case the potential for achievable speed-up is reduced but for 3-D problems we still might have enough parallelism compared with 2-D problems. However, this remains to be verified by numerical tests.

4 The acceleration parameters

As we noted earlier ADI can be efficient when we cycle with a decreasing sequence of parameters $\{\rho_i, i = 1, ..., l\}$. In fact for the model problem there is a complete theory on how to select the parameters optimally. This makes the algorithm converge within discretization errors, in a number of steps of the order of $O(n^{1/l})$ with a fixed l [15]. If $HV = VH$, and H and V are symmetric, positive definite, there exists satisfactory theory based on common eigenvectors [6]. If V and H commute but are not symmetric with possibly complex eigenvalues, G. Starke [20] derives an algorithm to determine the optimal ρ values for $l = 1$ and $l = 2$, and N. Ellner and E. Wachspress propose an algorithm for the case when the imaginary parts of eigenvalues are small relative to the real part. In all of the previous cases, these algorithms or formulas require some a-priori knowledge about the spectra of H and V. However, in practice the condition that H and V commute is much too restrictive. Actually, it dictates that the underlying PDE be separable in which case faster techniques may exist. In this paper we are interested in the more general situation where the PDE is not separable. This implies that there no longer exists a common set of eigenvectors, and this makes the analysis difficult. If H and V are symmetric, then one still could derive an *upper bound* for the spectral radius [23].* If H and V are not symmetric, which is true in the presence of *convection* terms in the underlying PDE, then the above upper bounds no longer hold.

As a result, for the general case, we must turn to heuristics to find optimal or nearly optimal iteration parameters ρ_i. For convenience we define ADI(k, l) to be the ADI(k) iteration using l parameters.

*Actually, for l greater than 1 a modified form of the classical ADI formulation is required.

4.1 Multigrid Motivation

We now go back to the simple case where $l = 1, HV = VH$, and H and V are symmetric, positive definite. Also assume that we know a, b such that

$$a \leq \lambda(H), \lambda(V) \leq b$$

where $\lambda(A)$ denotes the spectrum of the matrix A. Then the optimal ρ is given by $\sqrt{ab}$ [23]. For model problem $a \approx h^2, b \approx 2$, hence $\rho \approx \sqrt{2}h$. In other words the optimal ρ is linearly proportional to h. In general this would be the case when the *diffusion* term dominates in the discretized matrix (If h becomes very small, the discretized matrix is more dominated by diffusion terms).

Based on this observation we first find the optimal value of ρ in a much coarser grid, then we use the above relationship to predict the optimal ρ for the current grid. The following algorithm attempts to find an optimal set of l values, for a combination ADI-GMRES, starting from a coarser grid with t initial values. Note that these parameters are likely to be optimal only for the combination ADI-GMRES and not for ADI considered as a separate algorithm. The difference between the two can be quite important.

ALGORITHM **4.1.** *MG(n, t, k, l)*

1. *Given $N = n^2$ large enough, choose a coarser grid with $h_m = 4h$, and $N_m = N/16$.*

2. *Choose $\rho_1 \geq \rho_2 \geq \cdots \geq \rho_t \geq 0$, such that ρ_1 is large enough and ρ_t is small enough.(If the matrix is properly normalized, $\rho_1 \approx 0.1, \rho_t \approx 10, t \approx 6$ for model problem.)*

3. *Run $GMRES(m) - ADI(k)$ t times, each with ρ_i as its single parameter, until convergence.*

4. *Find ρ_j with the lowest iteration number.*

5. *Spread l values of ρ around $\frac{\rho_j}{4}$ (which is the estimated optimal ρ for the original grid)*

In step 1 we even could start from $h_m = 8h$, depending on the size of N. In step 3 rather than terminating when the residual norm is reduced by a given ϵ, we could alternately terminate with a fixed number of iterations and in step 4, find ρ_j with the smallest residual norm.

Regarding the cost of the procedure, it is clearly impossible to predict it exactly. An estimate may be obtained only by making the assumption that the number of iterations required to reduce the residual norm by a factor of ϵ is proportional to $\sqrt{N} = n$. Then the single run of GMRES(m)-ADI(k) in step 3 would be roughly $\frac{1}{16 \cdot 4}$ of the cost for the original size using the same ρ. Under this assumption, the cost of MG(n, t, k, l) would be 5/64 of the cost for the full system, when $t = 5$. For $h_m = 8h$ the cost becomes basically negligible.

The main feature of the algorithm is that by utilizing a multigrid motivated heuristic we avoid the otherwise sensitive problem of finding the optimal ρ, while by choosing a sequence of ρ parameters we expect to exploit the full power of ADI. Note that the heuristic is based on the assumption on a simple growth rule for the optimal parameters when h varies. We do not know whether this rule is valid for the general problems which we are considering. The heuristics can be still be used in these cases and seems to perform quite well, according to our numerical tests.

5 Numerical Experiments

We consider three test problems based on elliptic PDE's on square grids. We discretized the problems using centered finite difference discretizations in all cases. The mesh sizes vary from test to test and are reported independently in this section.

- **Problem 1** Poisson Equation on a Square

$$\begin{aligned} -\triangle u &= f, \ \Omega = [0,1] \times [0,1] \\ u &= 0 \ on \ \delta\Omega \\ f &= x(1-x) + y(1-y) \end{aligned}$$

- **Problem 2** Elman's problem [9]

$$\begin{aligned} -(bu_x)_x - (cu_y)_y + (du)_x + du_x + (eu)_y + eu_y + fu &= g \qquad (11) \\ \Omega = [0,1] \times [0,1] & \\ u = 0 \ on \ \delta\Omega & \end{aligned}$$

where $b = \exp(-xy)$, $c = \exp(xy)$, $d = \beta(x+y)$,
$e = \gamma(x+y)$, $f = \frac{1}{(1+xy)}$,
and g is such that exact solution $u = x\exp(xy)\sin(\pi x)\sin(\pi y)$

- **Problem 3** Discontinuous coefficients problem described by Eq. (11) where b and c are given by

$$\begin{aligned} b &= 100, \ 0 < x < 0.5 \\ &= 0.01, \ 0.5 < x < 1 \\ c &= 100, \ 0 < y < 0.5 \\ &= 0.01, \ 0.5 < y < 1 \end{aligned}$$

First we note that all of the test problems are in 2-D geometries. By a suitable reordering $H^{(k)}$ and $V^{(k)}$ can be made into matrices of bandwidth $2k+1$. Thus, using the horizontal natural ordering for $V^{(k)}$ and the vertical natural ordering for $H^{(k)}$, both matrices will be pentadiagonal when $k = 2$. A direct band solver is used for (9). Ignoring the initial factorization costs, the cost of solving (9) is roughly k times that of ADI(1), the classical ADI. In our experiments we used $k = 2, 4, 8$. For these values of k we expect the cost of solving (9) to remain reasonable. However, this may not be true for large problems from three-dimensional domains, since the bandwidth of (9) may become too large for banded solvers to be economical and we may have to consider *iterative* methods. As for the MG(n, t, k, l) we set t=5. We chose l to be 6, but depending on the problems the optimal l might vary.

A flexible GMRES routine allowing variable preconditioner at each iteration was used with m=10, $\epsilon = 10^{-6}$ for the *outer* iteration, where m is the dimension of Krylov subspace associated with the GMRES method. As for the finite element discretizations we used piecewise linear functions on triangles. In 2-D problems they give rise to sparse matrices with 7 diagonals. Experiments were done on a Cray-2 and a Sun-4. The timings reported were obtained on the Cray-2.

Table 1 contains the iteration number and CPU time on a Cray-2 of GMRES(10)-ADI(1,6) with a fixed parameter ρ taking various values versus those of GMRES(10)-ILU(0) and GMRES(10)-ILU(8). For the GMRES-ILU(0) and GMRES-ILUT(8) we used level-scheduling and jagged diagonal data structures for efficient vectorization on Cray-2[16]. For problem 3 with $\rho = 0.1$ and $\rho = 0.2$ the convergence was not reached within a reasonable amount of CPU time, so they are listed as 'N.A.'. For problem 2 with FDM, ADI(1,6) preconditioning with $\rho = 0.1, 0.2, 0.5$ takes less time than ILU(0) or ILUT(8) preconditioning. For problem 3 with FDM, ILU(0) or ILUT(8) preconditioning works better. Note that the MG(n, t, k, l) heuristic was not adopted. Also with GMRES-ILUT the parallelism on a one processor Cray-2 is limited to vectorization, the length of vector register, while for ADI(1,6) preconditioning the mimimum parallelism is n=128.(If we use cyclic reduction the parallelism could much exceed it.) So on massively parallel machines ADI preconditioning has a bigger potential for parallelism.

Table 2-4 compares the iteration number with the ρ parameters obtained by the MG$(n, 5, k, 6)$ heuristic versus the iteration numbers with the optimal ρ, l=1, obtained empirically. It shows that MG$(n, 5, k, 6)$ is always superior to the optimal ρ, l=1.

Table 5-9 lists for various but fixed ρ the iteration number of GMRES-ADI(k,6) for k=1, 2, 4. It is intended to show the sensitivity to ρ and the effectiveness of increasing k. The sensitivity of iteration numbers to ρ can be easily seen. We also see that the iteration number is decreasing as k increases. But considering the cost of ADI(k,6) to be roughly k times that of ADI(1,6) we need a drop in iteration number in the same fraction, to be competitive in total costs. However, the table shows that is rarely the case. In a quite few cases we even notice stagnation. One possible explanation is that the iteration number for ADI(1,6) is already quite small.

6 Conclusion

We have proposed a block version of the ADI algorithm based on a domain decomposition viewpoint. In addition, we have derived a simple yet effective way of getting optimal acceleration parameters for the ADI-preconditioned GMRES iteration, based on a multigrid approach. The numerical experiments reported indicate that the approach holds some promise for the parallel solution of elliptic PDE's on arbitrary domains. Two attractive features of the method are its ease of implementation, and its generality. Although a full comparison with the best domain decomposition approaches has yet to be made, the method itself can be viewed as a domain decomposition approach and for this reason its performance may be comparable. However, a distinct feature from a traditional domain decomposition approach is the use of acceleration parameters.

Acknowledgements. The authors would like to acknowledge the support of the Minnesota Supercomputer Institute which provided the computer facilities and an excellent research environment to conduct this research.

References

[1] E. C. Anderson, *Parallel implementation of preconditioned conjugate gradient methods for solving sparse systems of linear equations*, Technical Report 805, CSRD, University of Illinois, Urbana, IL, 1988. MS Thesis.

[2] E. C. Anderson and Y. Saad, *Solving sparse triangular systems on parallel computers*, International Journal of High Speed Computing, 1:73–96, 1989.

[3] O. Axelsson, *Conjugate gradient type-methods for unsymmetric and inconsistent systems of linear equations*, Technical Report 74-10, CERN, Geneva, 1974.

[4] O. Axelsson, *A survey of preconditioned iterative methods for linear systems of algebraic equations*, BIT, 25:166–187, 1985.

[5] D. Baxter, J. Saltz, M. H. Schultz, S. C. Eisenstat, and K. Crowley, *An experimental study of methods for parallel preconditioned Krylov methods*, Technical Report 629, Computer Science, Yale University, New Haven, CT, 1988.

[6] G. Birkhoff, R. Varga, S. R., and D. Young, *Alternating direction implicit methods*, in Advances in Computers, pp.189-273, Academic Press, New York, 1962

[7] J. Douglas, *Alternating direction methods for three space variables*, Numerische Mathematik Vol. 4, 1962, pp. 41-63.

[8] N. Ellner and E. Wachspress, *Alternating direction implicit iteration for systems with complex spectra*, SIAM J. Numerical Analysis, Vol. 28, No. 3, pp. 859-870, 1991

[9] H. ELMAN, *Iterative Methods for Large, Sparse, Nonsymmetric Systems of Linear Equations*, Ph. D Thesis, Yale University, 1982

[10] D. GANNON AND J. VAN ROSENDALE, *On the impact of communication complexity in the design of parallel algorithms*, IEEE Trans. Comp., C-33(12):1180–1194, 1984.

[11] S. L. JOHNSSON, *Solving narrow banded systems on ensemble architectures*, ACM, TOMS, 11(3), 1985.

[12] S. L. JOHNSSON, Y. SAAD, AND M. H. SCHULTZ, *The alternating direction algorithm on multiprocessors*, SIAM J. Sci. Statist. Comp, 8:686–700, 1987.

[13] W. J. LAYTON AND P. J. RABIER. *Peaceman Rachford procedure and domain decomposition for finite element problems*, Technical report, University of Pittsburgh, Pittsburgh, PA, (1991).

[14] J. A. MEIJERINK AND H. A. VAN DER VORST, *An iterative solution method for linear systems of which the coefficient matrix is a symmetric M-matrix*, Math. Comp., 31(137):148–162, 1977.

[15] D. PEACEMAN AND H. RACHFORD, *The numerical solution of elliptic and parabolic differential equations*, Journal of SIAM., Vol. 3, pp. 28-41, 1955

[16] Y. SAAD, *Krylov subspace methods on supercomputers*, SIAM J. Sci. Stat, Vol. 10, No. 6, pp. 1200-12332, Nov, 1989

[17] Y. SAAD, *ILUT: A Dual Threshold Incomplete LU Factorization*, UMSI 92/38, 1992

[18] Y. SAAD, *Highly parallel preconditioners for general sparse matrices*, Technical Report, University of Minnesota, Army High Performance Computing Research Center, Minneapolis, Minnesota, 1992.

[19] Y. SAAD, *ILUM: A parallel multi-elimination ILU preconditioner for general sparse matrices*, Technical Report, University of Minnesota, Army High Performance Computing Research Center, Minneapolis, Minnesota, 1992. In preparation.

[20] G. STARKE, *Optimal alternating direction implicit parameters for nonsymmetric systems of linear equations*, SIAM J. Numerical Analysis, Vol. 28, No. 5, pp. 1431-1445, 1991

[21] R. VARGA, *Matrix Iterative Analysis*, Prentice-Hall, New York, 1962

[22] E. WACHSPRESS, *Iterative Solution of Elliptic Systems*, Prentice-Hall, New York, 1966

[23] D. YOUNG, *Iterative Solution of Large Linear Systems*, Academic Press, New York, 1971.

[24] Z. ZLATEV, *Use of iterative refinement in the solution of sparse linear systems*, SIAM J. Numer. Anal., 19:381–399, 1982.

FDM Problem 2 with $\gamma = 50$, β=1								
N = 128^2	ρ=0.1	ρ=0.2	ρ=0.5	ρ=1	ρ=2	ρ=3	ILU(0)	ILUT(8)
Iter #	15	12	13	36	78	108	93	14
CPU	3.29	2.61	2.73	7.55	15.97	22.83	4.28	4.26
FDM Prob 3 with $\gamma = 50, \beta = 1$								
Iter #	N.A.	N.A.	63	36	49	43	191	42
CPU	N.A.	N.A.	13.41	10.28	10.25	9.17	7.46	6.45

Table 1: Iteration and CPU time of GMRES(10)-ADI(1,6) with constant ρ vs. GMRES(10)-ILU on Cray-2. N is the dimension of the matrix.

Problem 1		
	Heuristic	Optimal ρ
N=64^2		
k	l=6	l = 1
1	7	8
2	6	7
4	4	5
N=128^2		
1	8	12
2	7	9
4	5	7

Table 2: ADI(k,6) preconditioning with MG($n, 5, k, 6$) heuristic vs ADI(k,6) preconditioning with a theoretically determined optimal ρ

FDM Problem 2 $\gamma = 50, \beta = 1$		
	Heuristic	Optimal ρ
N=64^2		
k	l=6	l = 1
1	9	9
2	3	4
4	3	3
N=128^2		
1	9	12
2	5	6
4	3	5

Table 3: ADI(k,6) preconditioning with MG(n,5,k,6) heuristic vs ADI(k,6) preconditioning with a theoretically determined optimal ρ

FDM Problem 3 $\gamma = 50, \beta = 1$		
N=64^2		
k	Heuristic l=6	Optimal ρ l = 1
1	18	23
2	12	16
4	10	11
N=128^2		
1	32	35
2	17	22
4	13	15

Table 4: ADI(k,6) preconditioning with MG(n,5,k,6) heuristic vs ADI(k,6) preconditioning with a theoretically determined optimal ρ

Problem 1 with N=128^2										
	ρ									
k	0.05	0.1	0.15	0.2	0.25	0.3	0.35	0.4	0.45	0.5
1	13	14	12	14	20	24	29	29	32	35
2	10	9	9	9	10	12	12	14	18	19
4	7	7	8	9	10	11	12	13	17	19

Table 5: Iteration of GMRES(10)-ADI(k,6) for various values of ρ and k

FDM Problem 2 with $N = 64^2, \gamma = 50$, β=1										
	ρ									
k	0.05	0.1	0.15	0.2	0.25	0.3	0.35	0.4	0.45	0.5
1	25	15	14	13	12	11	10	10	9	9
2	8	5	4	4	4	4	4	5	5	5
4	5	3	3	3	4	4	4	5	5	5

Table 6: Iteration of GMRES(10)-ADI(k,6) for various values of ρ and k

FDM Problem 2 with $N = 128^2, \gamma = 50$, β=1										
	ρ									
k	0.05	0.1	0.15	0.2	0.25	0.3	0.35	0.4	0.45	0.5
1	19	15	13	12	12	12	12	12	12	13
2	8	6	6	7	8	9	10	11	14	17
4	5	5	6	7	8	9	10	11	13	16

Table 7: Iteration of GMRES(10)-ADI(k,6) for various values of ρ and k

FDM Problem 3 with $N = 64^2, \gamma = 50$, β=1									
	ρ								
k	2	3	4	5	6	7	8	9	10
1	25	26	25	25	24	23	23	23	25
2	18	18	17	16	16	17	19	20	22
4	11	11	12	14	16	18	19	20	22

Table 8: Iteration of GMRES(10)-ADI(k,6) for various values of ρ and k

FDM Problem 3 with $N = 128^2, \gamma = 50$, β=1									
	ρ								
k	1	1.5	2	2.5	3	3.5	4	4.5	5
1	49	48	48	45	43	38	35	35	36
2	26	24	23	22	23	24	26	27	29
4	16	15	16	18	19	22	27	30	33

Table 9: Iteration of GMRES(10)-ADI(k,6) for various values of ρ and k

A new application for generalized M–matrices

Reinhard Nabben *

Abstract. In the recent papers [7] and [13], the new classes of the so called generalized M–matrices and generalized H–matrices were introduced. These classes of block matrices are extensions of the well–known classes of M–matrices and H–matrices. They arise for example in the numerical solution of Euler equations and in the study of invariant tori of dynamical systems. Here, we give a new application for the theory of these classes of block matrices. We prove that the block matrices W, which arise in the numerical solution of the generalized Stokes equations using flux difference splittings are such that $W + W^H$ are generalized M–matrices. Furthermore, we establish some new theoretical results for generalized H–matrices, as for example the convergence of the SSOR-method, and we give a new equivalent condition for a matrix to be a generalized H–matrix, which complements the six known equivalent conditions given in [13].

Key words. generalized H–matrices, generalized M–matrices, block SOR–method, block SSOR–method, Euler equations, Stokes equations.

AMS (MOS) subject classifications. 65F10, 65N20, 15A48 .

1 Introduction and Notation

Elsner and Mehrmann discuss in [7] a special class of block matrices $A = [A_{ij}] \in \mathbb{C}^{km,km}$, where the blocks $A_{ij} \in \mathbb{C}^{k,k}$ are Hermitian and the off-diagonal blocks A_{ij}, $i \neq j$, are negative semidefinite. They called such a matrix a generalized Z–matrix. These block matrices arise for example in the numerical solution of Euler equations [11],[3] and in the study of invariant tori of dynamical systems [4]. In many cases, the diagonal blocks are positive definite. Then this class of matrices

*Institute for Computational Mathematics Kent, Ohio 44242. Permanent address: Universität Bielefeld, Fakultät für Mathematik, Postfach 100131, 4800 Bielefeld 1, FRG. email: nabben@mathematik.uni-bielefeld.de. Research supported by the Deutsche Forschungsgemeinschaft.

is denoted by

$\hat{Z}^k_m := \{A = [A_{ij}] \in \mathbb{C}^{mk,mk} | A_{ij} = A^H_{ij} \in \mathbb{C}^{k,k}, \quad A_{ii} > 0, A_{ij} \leq 0 \text{ for } i,j = 1,\ldots,m, i \neq j\}$,

where $A_{ii} > 0$ means that A_{ii} is positive definite. If a generalized Z–matrix satisfies that there exits a positive vector $u^T = [u_1, \ldots, u_m]$, such that the matrix $R_i(A,u) := \sum_{j=1}^m u_j A_{ij}$ is positive definite for $i = 1, \ldots, m$, then, A is called a generalized M–matrix. This set of matrices is denoted by:

$M^k_m := \{A \in \hat{Z}^k_m | \text{ there exists } u \in \mathbb{R}^m_+ \text{ with } \sum_{j=1}^m u_j A_{ij} > 0 \text{ for all } i \in < m >\}$,

where $< m > := \{1, 2, \ldots, m\}$. The condition $\sum_{j=1}^m u_j A_{ij} > 0$ is in some sense a block diagonal dominance criterion. Among other things in [13], this criterion is generalized to be independent of the sign of the off-diagonal blocks as in the point case, i.e., 1×1 blocks. This led in [13] to the classes D^k_m and H^k_m of block matrices:

$D^k_m := \{A = [A_{ij}] \in \mathbb{C}^{mk,mk} | A_{ij} = A^H_{ij} \in \mathbb{C}^{k,k} \text{ and } A_{ii} > 0 \text{ for } i,j \in < m >\}$;

$H^k_m := \{A \in D^k_m | \text{ there exists } u \in \mathbb{R}^m_+ \text{ with } u_i |A_{ii}| - \sum_{\substack{j=1 \\ j \neq i}}^{m} u_j |A_{ij}| > 0 \text{ for } i \in < m >\}$,

where $|A_{ij}| := (A_{ij} A_{ij})^{\frac{1}{2}}$. Thus, $A = [A_{ij}] \in D^k_m$ is a generalized H–matrix if its block comparison matrix $\mathcal{M}(\mathrm{A}) = [\tilde{A}_{ij}]$ with

$$\begin{aligned} \tilde{A}_{ii} &:= |A_{ii}| \quad \text{for} \quad i = 1, \ldots, m, \\ \tilde{A}_{ij} &:= -|A_{ij}| \quad \text{for} \quad i,j = 1, \ldots, m, \quad i \neq j, \end{aligned}$$

is a generalized M–matrix. Obviously each generalized M–matrix is a generalized H–matrix; hence $M^k_m \subseteq H^k_m$. If $k = 1$, i.e., the blocks are all 1×1 matrices, we obtain the well-known classes of Z–matrices, M–matrices, and H–matrices.

It is shown in [13] that the matrices M, which arise in the numerical solution of the Euler equations, satisfy $M + M^H \in M^k_m$ or $\mathcal{M}(M) + \mathcal{M}(M^H) \in M^k_m$. Therefore, this is the main assumption of the theorems in [7] and [13]. Here, we give a new application for the theory for generalized M–matrices and generalized H–matrices. We prove that a block matrix W, which arises in the numerical solution of the generalized Stokes problem using flux difference splittings [F], satisfies $W + W^H \in M^k_m$. Furthermore, we establish in the next section some new theoretical results for generalized H–matrices, as for example the convergence of the block SSOR-method, and we add a new equivalent condition for a matrix to be a generalized H–matrix, which complements the six given in [13]. Finally, we mention that the results of this paper and of [13] are part of [12].

2 Iterative methods for generalized H–matrices

In this section we establish convergence for iterative methods for generalized H–matrices.

Let $A = D - L - U$ be the block standard decomposition of A, where D is the block diagonal of A, $-L$ is the related lower triangular part of A and $-U$ is the related upper triangular part of A. Then, the JOR, SOR and SSOR iteration matrices are defined as

$$H_{JOR}(\omega) := (1-\omega)I + \omega D^{-1}(L+U), \tag{2.1}$$

$$H_{SOR}(\omega) := (D-\omega L)^{-1}((1-\omega)D + \omega U), \tag{2.2}$$

$$H_{SSOR}(\omega) := (D-\omega U)^{-1}[(1-\omega)D+\omega L](D-\omega L)^{-1}[(1-\omega)D+\omega U], \tag{2.3}$$

where I is the identity matrix.

The convergence results given in this section are based on the following Lemma, which is given in [13].

Lemma 2.1. *Let $A, M, N \in \mathbb{C}^{n,n}$ with $A = M - N$. If for all $t \in \mathbb{R}$*

$$A_t := M + M^H - (e^{it}N + e^{-it}N^H) > 0,$$

then $\rho(M^{-1}N) < 1$. If $A_t \geq 0$ for all $t \in \mathbb{R}$, then $\rho(M^{-1}N) \leq 1$.

Here, $\rho(M^{-1}N)$ denotes the spectral radius of $M^{-1}N$. An earlier version of this lemma, with the extra assumption that $M + M^H$ is positive definite, was given in [7]. However, this small extension will be useful in our subsequent proof of Theorem 2.3. Furthermore, we need the following Lemma, given in [14].

Lemma 2.2. *Let $A \in \mathbb{C}^{n,n}$ be Hermitian. Let $M, N \in \mathbb{C}^{n,n}$ with $A = M - N, M$ nonsingular and $M^H + N > 0$. Then, $\rho(M^{-1}N) < 1$ if and only if $A = M - N > 0$.*

Now, assume $A \in D^k_m$ and $\mathcal{M}(A) + \mathcal{M}(A^H) \in M^k_m$, i.e. there exists a positive vector $u \in \mathbb{R}^m$ such that

$$2u_iA_{ii} - \sum_{\substack{j=1\\j\neq i}}^{m} u_j(|A_{ij}| + |A_{ji}|) > 0 \quad \text{for all } i = 1, \ldots, m.$$

Then, we can define

$$s(A,u) := \max_{i\in<m>} \left\{ \rho(\frac{1}{2u_i}A_{ii}^{-1}\sum_{\substack{j=1\\j\neq i}}^{m} u_j(|A_{ij}| + |A_{ji}|)) \right\}. \tag{2.4}$$

Since, by assumption, $2u_iA_{ii} - \sum_{j\neq i} u_j(|A_{ij}| + |A_{ji}|)$ is positive definite and since $2u_iA_{ii} + \sum_{j\neq i} u_j(|A_{ij}| + |A_{ji}|)$ is the sum of a positive definite matrix and positive semidefinite matrices, we obtain with Lemma 2.2 that

$$s(A,u) < 1.$$

Using the value $s(A,u)$, we can extend Theorem 4.3 of [13] to

Theorem 2.3. *Let $A \in D_m^k$ with $\mathcal{M}(A) + \mathcal{M}(A^H) \in M_m^k$. Let $H_{JOR}(\omega)$ and $H_{SOR}(\omega)$ be as in (2.1), (2.2) and let $s(A,u)$ be as in (2.4). Then,*

$$\rho(H_{JOR}(\omega)) < 1, \quad \rho(H_{SOR}(\omega)) < 1, \quad \text{for all} \quad 0 < \omega < \frac{2}{1 + s(A,u)}.$$

Remark: Note that

$$1 < \frac{2}{1 + s(A,u)} \leq 2,$$

since $s(A,u) < 1$.

Proof. We only give the proof for the JOR method. The proof for the SOR method is analogous. In the following we show that for all $t \in \mathbb{R}$ the matrix A_t, defined by

$$A_t = \frac{2}{\omega} D - \left(e^{it}((\frac{1}{\omega} - 1)D + L + U) + e^{-it}((\frac{1}{\omega} - 1)D + L^H + U^H) \right)$$

is positive definite (i.e., $A_t > 0$). We obtain

$$\begin{aligned} A_t &= \sum_{\substack{s \neq j \\ s<j}} \begin{bmatrix} 0 & & & & 0 \\ & |A_{sj}| & & -e^{it} A_{sj} & \\ & & 0 & & \\ & -e^{-it} A_{sj} & & |A_{sj}| & \\ 0 & & & & 0 \end{bmatrix} \\ &+ \sum_{\substack{s \neq j \\ s<j}} \begin{bmatrix} 0 & & & & 0 \\ & |A_{js}| & & -e^{-it} A_{js} & \\ & & 0 & & \\ & -e^{it} A_{js} & & |A_{js}| & \\ 0 & & & & 0 \end{bmatrix} \\ &+ \begin{bmatrix} \ddots & & \\ & 2(\frac{1}{\omega} - cos(t)(\frac{1}{\omega} - 1))A_{ss} - \sum_{s \neq j}^{m} (|A_{sj}| + |A_{js}|) & \\ & & \ddots \end{bmatrix}. \end{aligned}$$

With Lemma 3.1 of [13], each of the first two terms are the sums of positive semi-definite matrices. Hence, it is easy to see that A_t is positive definite for all ω with $0 < \omega \leq 1$.

Now, let $1 < \omega < \frac{2}{1+s(A,u)}$. Then,

$$\frac{1}{\omega} - cos(t)(\frac{1}{\omega} - 1) \geq \frac{2 - \omega}{\omega}$$

for all $t \in \mathbb{R}$. Using Lemma 2.2, we obtain

$$2\frac{2 - \omega}{\omega} u_i A_{ii} - \sum_{\substack{j=1 \\ j \neq i}}^{m} u_j(|A_{ij}| + |A_{ji}|) > 0,$$

which is equivalent to

$$0 < \omega < \frac{2}{1 + \rho(\frac{1}{2u_i} A_{ii}^{-1} (\sum_{j \neq i} u_j (|A_{ij}| + |A_{ji}|)))}.$$

Thus, A_t is positive definite for all $t \in \mathbb{R}$ if

$$0 < \omega < \frac{2}{1 + s(A, u)}.$$

□

Theorem 2.3 states that overrelaxation is possible for the JOR-method and the SOR-method. Nevertheless, it is still an open question whether the number

$$\frac{2}{1 + \rho(D^{-1}(L + U))},$$

where D, L and U are block matrices, is an upper bound for the convergent block SOR parameter ω, which is a well-known result for the point case (see [15]). However, Theorem 2.3 seems to be a first step.

Now, let us look at the SSOR method with $\omega = 1$. Then, the iteration matrix $H_{SSOR}(1)$ becomes

$$H_{SSOR}(1) = (D - U)^{-1} L (D - L)^{-1} U.$$

We have

Theorem 2.4. *Let $A \in D_m^k$ and assume that there exists a positive vector $u \in \mathbb{R}^n$, such that*

$$u_i A_{ii} - \sum_{\substack{j=1 \\ j \neq i}}^{m} u_j (|A_{ij}| + |A_{ji}|) > 0 \quad \text{for all } i = 1, \ldots, m. \tag{2.5}$$

Then,

$$\rho((D - U)^{-1} L (D - L)^{-1} U) < 1.$$

Proof. In the following we consider the matrix $\hat{A} \in D_{2m}^k$ with

$$\hat{A} := \begin{bmatrix} D - L & -U \\ -L & D - U \end{bmatrix}.$$

We obtain

$$\hat{A} + \hat{A}^H = \begin{bmatrix} 2D - L - L^H & -U - L^H \\ -U^H - L & 2D - U - U^H \end{bmatrix} =$$

$$\left[\begin{array}{ccc|ccc}
\ddots & & & 0 & & \\
2A_{ii} & & A_{ij} & & \ddots & A_{ij}+A_{ji} \\
& \ddots & & & \ddots & \\
A_{ij} & & 2A_{jj} & & & \\
& & \ddots & 0 & & 0 \\
\hline
0 & & 0 & \ddots & & \\
& & & 2A_{ii} & & A_{ji} \\
& \ddots & & & \ddots & \\
A_{ij}+A_{ji} & & \ddots & A_{ji} & & 2A_{jj} \\
& & 0 & & & \ddots
\end{array}\right].$$

With (2.5) it follows that

$$u_iA_{ii} - \sum_{\substack{j=1\\j\neq i}}^{m} u_j|A_{ij}| \geq 0 \quad \text{and} \quad u_iA_{ii} - \sum_{\substack{j=1\\j\neq i}}^{m} u_j|A_{ji}| \geq 0 \quad \text{for all } i \in <m>.$$

Therefore, $\mathcal{M}(\hat{A}) + \mathcal{M}(\hat{A}^H) \in M_{2m}^k$. With Theorem 4.2 of [13], the splitting $\hat{A} = M - N$, given by

$$M := \begin{bmatrix} D-L & 0 \\ 0 & D-U \end{bmatrix}, \quad N := \begin{bmatrix} 0 & U \\ L & 0 \end{bmatrix},$$

is convergent, i.e., $\rho(M^{-1}N) < 1$. Thus, we have in particular for all $\hat{x}_0 = [x_0^H, x_0^H]^H, x_0 \in \mathbb{C}^{mk}$

$$\lim_{i\to\infty} (M^{-1}N)^i \hat{x}_0 = 0$$

(see [2] p.40). Since furthermore

$$M^{-1}N = \begin{bmatrix} 0 & (D-L)^{-1}U \\ (D-U)^{-1}L & 0 \end{bmatrix},$$

we obtain

$$\hat{x}_{i+2} = (M^{-1}N)^{i+2}\hat{x}_0 = (M^{-1}N)^2\hat{x}_i = \begin{bmatrix} (D-U)^{-1}L(D-L)^{-1}Ux_i \\ (D-L)^{-1}U(D-U)^{-1}Lx_i \end{bmatrix}.$$

Therefore, $\rho((D-U)^{-1}L(D-L)^{-1}U) < 1$. □

Note that assumption (2.5) of Theorem 2.4 states

$$\mathcal{M}\left(\tfrac{1}{2}D - L - U\right) + \mathcal{M}\left(\tfrac{1}{2}D - L^H - U^H\right) \in M_m^k$$

instead of $\mathcal{M}(\mathrm{A}) + \mathcal{M}(A^H) \in M_m^k$ as in Theorem 2.3. Nevertheless, Theorem 2.4 gives a convergence result for non-Hermitian block matrices.

Now we come to characterizations of generalized H–matrices. In [13], it is proved that the convergence of the Jacobi method (JOR method with $\omega = 1$), for a class of matrices which are associated with a given matrix $A \in D^k_m$, is a necessary and sufficient condition for A to satisfy $\mathcal{M}(\mathrm{A}) + \mathcal{M}(A^H) \in M^k_m$. Here, we establish the same result for the Gauss-Seidel method (SOR method with $\omega = 1$). Therefore, we need the following class of block matrices which is introduced in [13]:

Definition 2.5. *Let $A \in D^k_m$ and $u \in \mathbb{R}^m_+$. Then, $\Omega(A, u)$ is defined as*

$$\Omega(A,u) := \{B = [B_{ij}] \in D^k_m \mid \text{ there exists } v \in \mathbb{R}^m_+ \text{ such that for each } i \in <m>, \text{ there exists an } i_l \in <m> \text{ with } \sum_{\substack{j=1\\ j\neq i}}^m v_j(|B_{ij}| + |B_{ji}|) \leq \sum_{\substack{j=1\\ j\neq i_l}}^m u_j(|A_{i_l j}| + |A_{j i_l}|) \text{ and } v_i B_{ii} \geq u_{i_l} A_{i_l i_l}\}.$$

The class $\Omega(A, u)$ corresponds to the equimodular set of matrices associated with a point matrix A defined by Varga in [15].

The property of block generalized positive type is also a necessary and sufficient condition for a matrix $A \in D^k_m$ to be a generalized H–matrix [13]. Since we need this relation in the next section, we mention this property here too.

Definition 2.6. *Let $A = [A_{ij}] \in D^k_m$. A is of block generalized positive type if there exists a vector $u \in \mathbb{R}^m_+$ such that:*

i) $R_i(A,u) := u_i A_{ii} + \sum_{\substack{j=1\\ j\neq i}}^m u_j A_{ij} \geq 0$ *for all* $i \in <m>$,
and the set $J := \{i \in <m> \mid R_i(A,u) > 0\}$ *is nonempty.*

ii) *For each* $i_0 \in <m>$ *with* $i_0 \notin J$ *there exist indices* $i_1, \ldots, i_t \in <m>$ *with* $A_{i_l i_{l+1}} \neq 0, 0 \leq l \leq t-1$ *such that* $i_t \in J$.

Using these definitions, we obtain

Theorem 2.7. *Let $A \in D^k_m$. Then, the following conditions are equivalent:*

i) $\mathcal{M}(A) + \mathcal{M}(A^H)$ *is a generalized M–matrix.*

ii) *There exists a* $u \in \mathbb{R}^m_+$ *such that for each* $B \in \Omega(A,u)$

$$\rho(D_B^{-1}(L_B + U_B) < 1,$$

where $B = D_B - L_B - U_B$ *is the standard block decomposition of* B.

iii) *There exists a* $u \in \mathbb{R}^m_+$ *such that for each* $B \in \Omega(A,u)$

$$\rho((D_B - L_B)^{-1}U_B) < 1.$$

iv) $\mathcal{M}(A) + \mathcal{M}(A^H)$ *is of block generalized positive type.*

Proof. The parts $i) \Leftrightarrow ii)$ and $i) \Leftrightarrow iv)$ are proved in [13], and $i) \Rightarrow iii)$ follows directly from Theorem 2.3. For $iii) \Rightarrow i)$, consider the matrix

$$B = \begin{bmatrix} u_i A_{ii} & -\frac{1}{2}\sum_{j\neq i} u_j(|A_{ij}|+|A_{ji}|) & & & \\ -\frac{1}{2}\sum_{j\neq i} u_j(|A_{ij}|+|A_{ji}|) & u_i A_{ii} & & 0 & \\ & & u_i A_{ii} & & \\ & 0 & & \ddots & \\ & & & & u_i A_{ii} \end{bmatrix}.$$

Since $A \in \Omega(A,u)$, B is also in $\Omega(A,u)$ for each $i \in <m>$. Thus, we have

$$\rho((D_B - L_B)^{-1} L_B) < 1.$$

On the other hand, B is a block tridiagonal matrix, and therefore, B is a consistently ordered 2-cyclic matrix (see Varga [16] p. 102). But then, by using the block version of the Stein-Rosenberg Theorem (Corollary 5.13 of [7]), we obtain one of the following mutually exclusive relations:

$$\begin{aligned} \rho(D_B^{-1}(L_B+U_B)) = \rho((D_B - L_B)^{-1} L_B) &= 0 \quad \text{or} \\ \rho((D_B - L_B)^{-1} L_B) < \rho(D_B^{-1}(L_B+U_B)) &< 1. \end{aligned}$$

Thus, $\rho(D_B^{-1}(L_B+U_B)) < 1$. But since,

$$D_B^{-1}(L_B+U_B) = \begin{bmatrix} 0 & \frac{1}{2u_i} A_{ii}^{-1} \sum_{\substack{j=1\\ j\neq i}}^{m} u_j(|A_{ij}|+|A_{ji}|) & & & \\ \frac{1}{2u_i} A_{ii}^{-1} \sum_{\substack{j=1\\ j\neq i}}^{m} u_j(|A_{ij}|+|A_{ji}|) & 0 & & & \\ & & 0 & & \\ & & & \ddots & \\ & & & & 0 \end{bmatrix}$$

we obtain, following the proof given in [13] for part ii) $\Rightarrow$ i),

$$\rho\left(\frac{1}{2u_i} A_{ii}^{-1} \sum_{\substack{j=1\\ j\neq i}}^{m} u_j(|A_{ij}|+|A_{ji}|)\right) < 1.$$

Furthermore, we have

$$2u_i A_{ii} + \sum_{\substack{j=1\\ j\neq i}}^{m} u_j(|A_{ij}|+|A_{ji}|) > 0.$$

Using Lemma 2.2, it follows

$$2u_i A_{ii} - \sum_{\substack{j=1\\ j\neq i}}^{m} u_j(|A_{ij}|+|A_{ji}|) > 0 \text{ for } i \in <m>.$$

Hence, $\mathcal{M}(A) + \mathcal{M}(A^H)$ is a generalized M–matrix. □

With this new condition given here and those given in [13], there exist now six conditions for a block matrix $A \in D_m^k$ to satisfy condition i) of Theorem 2.7, namely that $\mathcal{M}(A) + \mathcal{M}(A^H)$ is a generalized M–matrix. In the case $k = 1$, i.e. the classical case, there are many more known equivalent conditions, which are necessary and sufficient for a Z–matrix to be an M–matrix, e.g. [1], and some more equivalent conditions for H–matrices [15]. However, Elsner and Mehrmann mention in [7] that some of these conditions do not carry over to our block case.

3 Generalized M–matrices in the numerical solution of Stokes equations

In this section, we give a new application for the theory of generalized M–matrices and generalized H–matrices. We discuss a special class of these matrices, which arise as coefficient matrices in the numerical solution of the generalized Stokes problem using flux difference splittings [F]. These matrices have the following form:

$$W := \begin{bmatrix} T_1 & S_1^- & & \\ S_2^+ & T_2 & \ddots & \\ & \ddots & \ddots & S_{p-1}^- \\ & & S_p^+ & T_p \end{bmatrix} \in \mathbb{C}^{p\cdot r\cdot k, p\cdot r\cdot k}, \tag{3.1}$$

where $T_j, S_j^-, S_j^+ \in \mathbb{C}^{r\cdot k, r\cdot k}$, $(j = 1, \ldots, p)$ are defined by

$$T_j := \begin{bmatrix} D_{1j} & -A_{\frac{3}{2},j}^- & & & \\ \ddots & \ddots & \ddots & & \\ & -A_{i-\frac{1}{2},j}^+ & D_{ij} & -A_{i+\frac{1}{2},j}^- & \\ & & \ddots & \ddots & \ddots \\ & & & -A_{r-\frac{1}{2},j}^+ & D_{rj} \end{bmatrix},$$

$$S_j^- := \begin{bmatrix} -B_{1,j+\frac{1}{2}}^- & & & & \\ & \ddots & & & \\ & & -B_{i,j+\frac{1}{2}}^- & & \\ & & & \ddots & \\ & & & & -B_{r,j+\frac{1}{2}}^- \end{bmatrix}, \tag{3.2}$$

$$S_j^+ := \begin{bmatrix} -B_{1,j-\frac{1}{2}}^+ & & & & \\ & \ddots & & & \\ & & -B_{i,j-\frac{1}{2}}^+ & & \\ & & & \ddots & \\ & & & & -B_{r,j-\frac{1}{2}}^+ \end{bmatrix}.$$

Here for $i = 0, \ldots, r$ and $j = 0, \ldots, p$

$A_{i+\frac{1}{2},j} = A_{i+\frac{1}{2},j}^+ - A_{i+\frac{1}{2},j}^- \in \mathbb{C}^{k,k}$ and $B_{i,j+\frac{1}{2}} = B_{i,j+\frac{1}{2}}^+ - B_{i,j+\frac{1}{2}}^- \in \mathbb{C}^{k,k}$

are decompositions of symmetric matrices $A_{i+\frac{1}{2},j}$ and $B_{i,j+\frac{1}{2}}$ into positive semidefinite parts and negative semidefinite parts. The diagonal blocks D_{ij} ($i = 1, \ldots, r$ and $j = 1, \ldots, p$) are given by

$$\begin{aligned} D_{ij} &= \frac{1}{2}(A_{i+\frac{1}{2},j}^+ + A_{i+\frac{1}{2},j}^- + A_{i-\frac{1}{2},j}^+ + A_{i-\frac{1}{2},j}^-) \\ &+ \frac{1}{2}(B_{i,j+\frac{1}{2}}^+ + B_{i,j+\frac{1}{2}}^- + B_{i,j-\frac{1}{2}}^+ + B_{i,j-\frac{1}{2}}^-). \end{aligned}$$

Furthermore,

$$A_{i+\frac{1}{2},j}^+ + A_{i+\frac{1}{2},j}^- > 0, \quad B_{i,j+\frac{1}{2}}^+ + B_{i,j+\frac{1}{2}}^- > 0. \tag{3.3}$$

The matrix W has the same block structure as the matrices which arise in the numerical solution of the Euler equations [11]. Moreover, if we have in this abstract formulation above that

$$\begin{aligned} A_{i+\frac{1}{2},j}^+ &= A^+, \quad A_{i+\frac{1}{2},j}^- = A^-, \\ B_{i,j+\frac{1}{2}}^+ &= B^+, \quad B_{i,j+\frac{1}{2}}^- = B^-, \end{aligned}$$

for $i = 0, \ldots, r$ and $j = 0, \ldots, p$, the matrix W is equal to the matrix M, which is discussed in [7] and [13]:

$$M := \begin{bmatrix} T_1 & S_1 & & \\ S_2 & T_1 & \ddots & \\ & \ddots & \ddots & S_1 \\ & & S_2 & T_1 \end{bmatrix} \in \mathbb{C}^{p\cdot r\cdot k, p\cdot r\cdot k}, \tag{3.4}$$

where $T_1, S_1, S_2 \in \mathbb{C}^{r\cdot k, r\cdot k}$ are defined by

$$T_1 := \begin{bmatrix} C & -A^- & & \\ -A^+ & C & \ddots & \\ & \ddots & \ddots & -A^- \\ & & -A^+ & C \end{bmatrix},$$

$$S_2 := \begin{bmatrix} -B^+ & & \\ & \ddots & \\ & & -B^+ \end{bmatrix}, \tag{3.5}$$

$$S_1 := \begin{bmatrix} -B^- & & \\ & \ddots & \\ & & -B^- \end{bmatrix}.$$

This means, in this abstract formulation above, that M is a special case of W. However, the special matrices with the physical coefficients used in [11] (M) and [F] (W) are different.

Since in general no block of W is equal to another block, the techniques of the proofs of the results, given in [7] for M, are not applicable to W. But similar to [13], we can show that

$$W + W^H \in M_{pr}^k \quad \text{and} \quad W + W^H \in M_p^{rk},$$

using the property of block generalized positive type:

Theorem 3.1. *Let W be as in* (3.1) *and* (3.2). *Then,*

$$W + W^H \in M_{pr}^k \quad \textit{and} \quad W + W^H \in M_p^{rk}.$$

Proof. The proof is similar to the proof given in [13] for the matrix M. We start with the partition (3.1).

Using (3.3), we obtain $T_j + T_j^H \in M_m^k$, with the weighting vector $u = [1, \ldots, 1]^T$, for all $j = 1 \ldots, p$. Thus, with Corollary 3.7 of [7], $T_j + T_j^H$ is positive definite. Therefore, $W + W^H \in \hat{Z}_p^{rk}$. Now, consider the first and last block row of $W + W^H$. We have

$$T_1 + T_1^H + S_1^- + S_2^+ > 0 \quad \text{and} \quad T_p + T_p^H + S_p^+ + S_{p-1}^- > 0. \tag{3.6}$$

Thus, considering the block graph of A, the matrix $W + W^H$ is of block generalized positive type.

Now, let us look at the partitioning (3.2). Obviously, we have $W + W^H \in \hat{Z}_{pr}^k$. Furthermore, the matrices given in (3.6) satisfy

$$T_1 + T_1^H + S_1^- + S_2^+ \in M_r^k \quad \text{and} \quad T_p + T_p^H + S_p^+ + S_{p-1}^- \in M_r^k$$

with the weighting vector $u = [1, \ldots, 1]^T$. Therefore, using (3.3) and the block graph of $W + W^H$, the matrix $W + W^H$ is also of block generalized positive type for the partitioning (3.2). With Theorem 2.7 we obtain

$$W + W^H \in M_{pr}^k \quad \text{and} \quad W + W^H \in M_p^{rk}.$$

□

Thus, the matrix W satisfies the main assumption of the theorems above and of [7] and [13] that $W + W^H \in M_{pr}^k$ and $W + W^H \in M_p^{rk}$. Therefore, all of the results given are applicable to the special matrix W. In particular, the block Jacobi method converges. Obviously the matrix W has the same structure as M, and both are consistently ordered 2-cyclic (see Varga [16]). Thus, the theorems given in [7] and [13] for the consistently ordered 2-cyclic generalized Z–matrix M are also true for W.

Furthermore, in the following we mention one further property; the convergence of the SSOR method for all ω with $0 < \omega < 2$.

The relation between the eigenvalues of the Jacobi method and those of the SSOR iteration matrix is given for consistenly ordered p-cyclic matrices [16] by the following theorem due to Varga, Niethammer, and Cai [17]:

Theorem 3.2. *Let $A \in \mathbb{C}^{mk,mk}$ be consistently ordered and p-cyclic. Let $H_{SSOR}(\omega)$ be as in* (2.3) *and $0 < \omega < 2$. If λ is an eigenvalue of $H_{SSOR}(\omega)$ with $\lambda \neq (1-\omega)^2$ and if δ satisfies*

$$[\lambda - (1-\omega)^2]^p = \lambda[\lambda + 1 - \omega]^{p-2}(2-\omega)^2\omega^p\delta^p, \tag{3.7}$$

then $\delta \in \sigma(D^{-1}(L+U))$. Conversly, if δ is an eigenvalue of $D^{-1}(L+U)$ and if λ satisfies (3.7) *with $\lambda \neq (1-\omega)^2$, then λ is an eigenvalue of $H_{SSOR}(\omega)$.*

Here, $\sigma(D^{-1}(L+U))$ denotes the spectrum of $D^{-1}(L+U)$. Note that Hadjidimos and Neumann [10] have remarked that

$$\lambda = (1-\omega)^2 \in \sigma(H_{SSOR}(\omega)) \quad \text{if and only if} \quad \delta = 0 \in \sigma(D^{-1}(L+U)).$$

As mentioned in [17], we obtain for $p = 2$

$$[\lambda - (1-\omega)^2]^2 = \lambda(2-\omega)^2\omega^2\delta^2. \tag{3.8}$$

Setting, $\hat{\omega} = \omega(2-\omega)$, it follows that

$$[\lambda + \hat{\omega} - 1]^2 = \lambda\hat{\omega}^2\delta^2. \tag{3.9}$$

Using equation (3.9), we establish in the following theorem that, for a given matrix A, the SSOR method converges for all $0 < \omega < 2$ if A is consistently ordered 2-cyclic and the Jacobi method converges:

Theorem 3.3. *Let $A \in \mathbb{C}^{mk,mk}$ be consistently ordered 2-cyclic. Let $H_{SSOR}(\omega)$ be as in* (2.3) *and let $\rho(D^{-1}(L+U)) < 1$. Then,*

$$\rho(H_{SSOR}(\omega)) < 1$$

for all ω with $0 < \omega < 2$

Proof. For $0 < \omega < 2$ we obtain $0 < \hat{\omega} \leq 1$, where $\hat{\omega} = \omega(2-\omega)$. Let $\mu \in \sigma(H_{SSOR}(\omega))$. Then we have:

$$(|\mu| - (1-\hat{\omega}))^2 \leq |\mu - (1-\hat{\omega})|^2 = |\mu + \hat{\omega} - 1|^2 = |\mu|\hat{\omega}^2|\delta|^2 \leq |\mu|\hat{\omega}^2\rho^2,$$

where $\delta \in \sigma(D^{-1}(L+U))$. Thus,

$$(|\mu| - (1-\hat{\omega}))^2 \leq |\mu|\hat{\omega}^2\rho^2,$$

and by elementary considerations we get $|\mu| < 1$. Thus, $\rho(H_{SSOR}(\omega)) < 1$ for $0 < \omega < 2$. □

Theorem 3.3 is not a new result. It follows from the works of Ehrlich [6] and D'Sylva and Miles [5]. However, they do not mention this result explicitly. Furthermore, by using (3.7), the above proof seems to be easier.

Using Theorem 3.3, we can summarize our results and the results of [7] and [13] for W and hence for M in the following

Corollary 3.4. *Let W be as in* (3.1), (3.2). *For both decompositions* (3.1) *and* (3.2) *of W we have:*

i) *the block Jacobi method converges;*
ii) *the block SOR method converges if* $0 < \omega < \frac{2}{1+\rho}$, *where* ρ *is the spectral radius of the related block Jacobi iterative matrix.*
iii) *the block SSOR method converges for all* ω *with* $0 < \omega < 2$.

Proof. The part *i*) follows from Theorem 2.7 and Theorem 3.1. Using *i*), the part *ii*) follows from Theorem 5.14 of [7]. Then, *iii*) follows from Theorem 3.3, since W is consistently ordered 2-cyclic. □

Acknowledgement. The author thanks C. Frohn for showing him the matrices arising in the solution of the Stokes equations.

References

[1] A. Berman and R.J. Plemmons, *Nonnegative Matrices in the Mathematical Sciences,* Academic Press, New York (1979).

[2] W.Bunse and A. Bunse-Gerstner, *Numerische Lineare Algebra*, B.G. Teubner, Stuttgart (1985).

[3] E. Dick and J. Linden, *A multigrid flux-difference splitting method for steady incompressible Navier-Stokes equations*, Proceedings of the GAMM Conference on Numerical Methods in Fluid Mechanics, Delft, September 1989.

[4] L. Dieci and J. Lorenz, *Block M–matrices and Computation of Invariant Tori,* SIAM J. Sci. Stat. Comput. 13 (1992), 885-903.

[5] E. D'Sylva and G.A. Miles, *The S.S.O.R. iteration scheme for equations with* σ_1 *ordering*, Computer Journal 6 (1964), 366-367.

[6] L.W. Ehrlich, *The block symmetric successive overrelaxation method*, J. Soc. Indust. Appl. Math. 12 (1964), 807-826.

[7] L. Elsner and V. Mehrmann, *Convergence of block iterative methods for linear systems arising in the numerical solution of Euler equations*, Numerische Mathematik 59 (1991), 541-559.

[8] C. Frohn, *Flux-Splitting-Methoden und Mehrgitterverfahren für hyperbolische Systeme mit Beispielen aus der Strömungsmechanik*, Dissertation, Universität Düsseldorf (1992).

[9] G.H. Golub and C.F. van Loan, *Matrix Computations*, Second Edition, The Johns Hopkins University Press, Baltimore and London (1989).

[10] A. Hadjidimos, M. Neumann, *Precise domains of convergence for the block SSOR Method associated with p-cyclic matrices*, BIT 29 (1989), 311-320.

[11] P.W. Hemker and S.P. Spekreijse, *Multiple Grid and Osher's Scheme for the efficient solution of the steady Euler equations*, Appl. Numer. Math. 2 (1986), 475-493.

[12] R. Nabben, *Konvergente Iterationsverfahren für unsymmetrische Blockmatrizen*, Dissertation, Universität Bielefeld (1991).

[13] R. Nabben, *On a class of matrices which arise in the numerical solution of Euler equations*, to appear in Numerische Mathematik.

[14] J.M. Ortega, R.J. Plemmons, *Extensions of the Ostrowski - Reich Theorem for SOR iterations*, Linear Algebra and its Applications 88/89 (1987), 559-573.

[15] R.S. Varga, *On recurring theorems on diagonal dominance*, Lin. Alg. Appl. 13 (1976), 1-9.

[16] R.S. Varga, *Matrix Iterative Analysis*, Prentice-Hall, Engelwood Cliffs, N.J. (1962).

[17] R.S. Varga and W. Niethammer, D.Y. Cai, *p-cyclic matrices and the symmetric successive overrelaxation method*, Lin. Alg. Appl. 58 (1984), 425-439.

On Symmetric Ultrametric Matrices

Richard S. Varga and Reinhard Nabben†*

Abstract. In a recent paper by S. Martínez, G. Michon, and J. San Martín [1], it was shown that if $A := [a_{i,j}]$ in $\mathbb{R}^{n\times n}$ is a symmetric strictly ultrametric matrix, then its inverse $A^{-1} := [\alpha_{i,j}]$ in $\mathbb{R}^{n\times n}$ is a strictly diagonally dominant Stieltjes matrix, with the additional property that

$$a_{i,j} = 0 \text{ if and only if } \alpha_{i,j} = 0.$$

Here, a generalization of this result to symmetric ultrametric matrices is given.

Key words. Stieltjes matrix, strictly ultrametric matrices.

AMS(MOS) subject classification. 15A57, 15A48.

1 Introduction

In a recent paper, S. Martínez, G. Michon, and J.San Martín [1] gave the following definition. For notation, let $N := \{1, 2, \cdots, n\}$ for any positive integer n.

Definition 1.1. *A matrix $A = [a_{i,j}]$ in $\mathbb{R}^{n\times n}$ is a symmetric strictly ultrametric matrix if*

$$\begin{array}{rl} i) & A \text{ is symmetric with nonnegative entries;} \\ ii) & a_{i,j} \geq \min\{a_{i,k}; a_{k,j}\} \text{ for all } i,k,j \text{ in } N; \\ iii) & a_{i,i} > \max\{a_{i,k} : k \in N\backslash\{i\}\} \text{ for all } i \in N, \end{array} \quad (1)$$

where, if $n = 1$, then (1iii) is interpreted as $a_{1,1} > 0$.

The result of [1, Theorem 1] is

*Institute for Computational Mathematics, Kent State University, Kent, OH 44242, USA.
The research of this author was partially supported by the National Science Foundation.
†Fakultät für Mathematik, Universität Bielefeld, D-4800 Bielefeld 1, FRG.
The research of this author was partially supported by the Deutsche Forsungsgemeinschaft.

Numerical Linear Algebra

Theorem 1.1. ([1]). *If $A = [a_{i,}]$ in $\mathbb{R}^{n\times n}$ is a symmetric strictly ultrametric matrix, then A is nonsingular and its inverse, $A^{-1} := [\alpha_{i,j}]$ in $\mathbb{R}^{n\times n}$, is a strictly diagonally dominant Stieltjes matrix (i.e., $\alpha_{i,j} \leq 0$ for all $i \neq j$ and $\alpha_{i,i} > \sum_{\substack{k=1\\k\neq i}}^{n} |\alpha_{i,k}|$ for all $1 \leq i, j \leq n$), with the additional property that*

$$a_{i,j} = 0 \;\; if \; and \; only \; if \;\; \alpha_{i,j} = 0. \tag{2}$$

For a shorter linear algebra proof of Theorem 1.1, see Nabben and Varga [2].

The result from Theorem 1, that the inverse of a symmetric strictly ultrametric matrix is a *strictly* diagonally dominant Stieltjes matrix, suggested that a possible weakening of the hypotheses of Definition 1.1 might be possible. Our modest goal here is to specifically weaken (1.*iii*) of Definition 1.1 to obtain a generalization, Theorem 2.2 below, of Theorem 1.1.

2 Main Result

To begin, we first state the following results of Nabben and Varga [2], which will be used in our constructions below. For additional notation, set

$$\xi_n := (1, 1, \cdots, 1)^T \text{in } \mathbb{R}^n.$$

Proposition 2.1. ([2]). *Let $A = [a_{i,j}]$ in $\mathbb{R}^{n\times n}$ be symmetric with all its entries nonnegative, and set*

$$\tau(A) := \min\{a_{i,j} : i, j \in N\}. \tag{3}$$

If $n > 1$, then A is a symmetric strictly ultrametric matrix if and only if $A - \tau(A)\xi_n\xi_n^T$ is completely reducible, i.e., there exist a positive integer r with $1 \leq r < n$ and a permutation matrix P in $\mathbb{R}^{n\times n}$ such that

$$P\left[A - \tau(A)\xi_n\xi_n^T\right]P^T = \begin{bmatrix} C & O \\ O & D \end{bmatrix}, \tag{4}$$

where $C \in \mathbb{R}^{r\times r}$ and $D \in \mathbb{R}^{(n-r)\times(n-r)}$ are each a symmetric strictly ultrametric matrix.

Theorem 2.1. ([2]). *Given any symmetric strictly ultrametric matrix A in $\mathbb{R}^{n\times n}$ ($n \geq 1$), there is an associated rooted tree for $N = \{1, 2, \cdots, n\}$, consisting of $2n - 1$, vertices, such that*

$$A = \sum_{\ell=1}^{2n-1} \tau_\ell \, \mathbf{u}_\ell \, \mathbf{u}_\ell^T, \tag{5}$$

where the vectors $\mathbf{u}_\ell$ *in* (5), *determined from the vertices of the tree, are nonzero vectors in* $\mathbb{R}^n$ *having only* 0 *and* 1 *components, and where the* τ_ℓ*'s in* (5) *are nonnegative and, with the notation that*

$$\chi(\mathbf{u}_\ell) := \quad \textit{sum of the components of} \quad \mathbf{u}_\ell, \tag{6}$$

satisfy the property that $\tau_\ell > 0$ *when* $\chi(\mathbf{u}_\ell) = 1$. *Conversely, given any rooted tree for* $N = \{1, 2, \cdots, n\}$ *with* $\tau_\ell > 0$ *when* $\chi(\mathbf{u}_\ell) = 1$, *then* $\sum_{\ell=1}^{2n-1} \tau_\ell \, \mathbf{u}_\ell \, \mathbf{u}_\ell^T$ *is a symmetric strictly ultrametric matrix in* $\mathbb{R}^{n\times n}$.

To generalize Theorem 1.1, (1*iii*) of Theorem 1.1 is weakened to allow for the case of equality.

Definition 2.1. *A matrix* $A = [a_{i,j}]$ *in* $\mathbb{R}^{n\times n}$ *is a symmetric pre-ultrametric matrix if*

$$\begin{aligned} &i) \quad A \ \textit{ is symmetric with nonnegative entries;} \\ &ii) \quad a_{i,j} \geq \min\{a_{i,k}; a_{k,j}\} \ \textit{ for all } \ i,k,j \ \in \ N; \\ &iii) \quad a_{i,i} \geq \max\{a_{i,k} : k \in N\backslash\{i\}\} \ \textit{ for all } \ i \in N. \end{aligned} \tag{7}$$

It is evident that a symmetric pre-ultrametric matrix *can* be singular, as choosing $A = O$ shows. Now, it easily follows from Theorem 2.1 that A is a symmetric pre-ultrametric matrix if and only if the representation of (5) is valid where the τ_ℓ's in (5) are just nonnegative numbers (i.e., no further restrictions on the τ_ℓ's are necessary). But this shows that if A is a symmetric pre-ultrametric matrix, then for each $\epsilon > 0$,

$$A(\epsilon) := A + \epsilon I_n \text{ (where } I_n \text{ is the identity matrix in } \mathbb{R}^{n\times n}) \tag{8}$$

is a symmetric strictly ultrametric matrix which, from Theorem 1.1, is necessarily nonsingular. Then on applying (5) of Theorem 2.1, $A(\epsilon)$ can be represented as

$$A(\epsilon) = \sum_{\ell=1}^{2n-1} \tau_\ell(\epsilon)\mathbf{u}_\ell\mathbf{u}_\ell^T \quad (\epsilon > 0), \tag{9}$$

where it is important to note that the vectors $\{\mathbf{u}_\ell\}_{\ell=1}^{2n-1}$ in (9) are *independent* of ϵ. (This is a consequence of the fact that the complete reduction steps of (4) of Proposition 2.1, which are applied to principal submatrices of $A(\epsilon)$ to obtain the representation of (5), yield vectors $\mathbf{u}_\ell$ which depend only on the vertices of the associated tree and are *independent* of ϵ. This will also be illustrated in the example in Section 2.)

This brings us to our next

Definition 2.2. *A matrix* $A = [a_{i,j}]$ *in* $\mathbb{R}^{n\times n}$ *is a symmetric ultrametric matrix if* A *is a symmetric pre-ultrametric matrix and if, from the representation* (9), *the vectors* $\{\mathbf{u}_\ell\}_{\ell=1}^{2n-1}$ *satisfy*

$$(iv) \text{ span } \{\mathbf{u}_\ell : \tau_\ell(0) > 0\} = \mathbb{C}^n, \tag{10}$$

where $\tau_\ell(0) := \lim_{\epsilon\to 0} \tau_\ell(\epsilon)$.

To couple Definition 2.1 with Definition 1.1, let $\{\mathbf{e}_j\}_{n=1}^n$ denote the set of unit basis vectors in $\mathbb{R}^n$ (i.e., $\mathbf{e}_1 = (1, 0, \cdots, 0)^T$, $\mathbf{e}_2 = (0, 1, 0, \cdots, 0)^T$, etc.). If A is a symmetric strictly ultrametric matrix in $\mathbb{R}^{n\times n}$, then from Theorem 2.1, each $\mathbf{e}_j$ $(1 \le j \le n)$ is some $\mathbf{u}_\ell$ in $\{u_\ell\}_{\ell=1}^{2n-1}$ and its associated multiplier, τ_ℓ in (5), is necessarily positive. Consequently, as (10) is then obviously satisfied, each symmetric strictly ultrametric matrix in $\mathbb{R}^{n\times n}$ is necessarily a symmetric ultrametric matrix.

We next establish

Lemma 2.1. *Assume that $A = [a_{i,j}]$ in $\mathbb{R}^{n\times n}$ is a symmetric pre-ultrametric matrix. Then, A is positive definite if and only if (10) is valid, i.e, if and only if A is a symmetric ultrametric matrix.*

Proof. Since A is by hypothesis a symmetric pre-ultrametric matrix, then on letting $\epsilon \to 0$, it follows from (9) and (10) that

$$A = \sum_{\ell=1}^{m} \tau_\ell(0)\ \mathbf{u}_\ell\ \mathbf{u}_\ell^T \text{ where } \tau_\ell(0) > 0 \text{ for all } 1 \le \ell \le m, \tag{11}$$

where $m \le 2n - 1$. (This amounts to throwing out those $\tau_\ell(0)$ in (9) which are zero, and then renumbering the remaining terms in (9)). Then for any $\mathbf{x} \in \mathbb{R}^n$,

$$\mathbf{x}^H A\mathbf{x} = \sum_{\ell=1}^{m} \tau_\ell(0)\,(\mathbf{x}^H \mathbf{u}_\ell)\,(\mathbf{u}_\ell^T \mathbf{x}) = \sum_{\ell=1}^{m} \tau_\ell(0) |\mathbf{u}_\ell^T \mathbf{x}|^2 \ge 0, \tag{12}$$

so that A is at least positive semi-definite. Equality holds in (12) (since $\tau_\ell(0) > 0$ for all $1 \le \ell \le m$) only if $\mathbf{u}_\ell^T \mathbf{x} = 0$ for all $1 \le \ell \le m$. Thus, $\mathbf{x}$ is orthogonal to every linear combination of the $\mathbf{u}_\ell$'s, i.e.,

$$\mathbf{x} \perp \text{span } \{\mathbf{u}_\ell : \tau_\ell(0) > 0\}.$$

If (10) is valid, then equality in (12) implies $\mathbf{x} = \mathbf{0}$ and $\mathbf{y}^H A\mathbf{y} > 0$ for all $\mathbf{y} \neq \mathbf{0}$ in $\mathbb{C}^n$, i.e., A is positive definite. Conversely, if (10) is not true, there is an $\mathbf{x} \neq \mathbf{0}$ with $\mathbf{x}^H A\mathbf{x} = \mathbf{0}$, so that A is singular. □

This brings us to our main result.

Theorem 2.2. *Let $A = [a_{i,j}]$ in $\mathbb{R}^{n\times n}$ be a symmetric ultrametric matrix, in the sense of Definition 2.1. Then, A is positive definite and its inverse, $A^{-1} := [\alpha_{i,j}]$ in $\mathbb{R}^{n\times n}$, is a diagonally dominant Stieltjes matrix, i.e., $\alpha_{i,j} \le 0$ for all $i \neq j$ and*

$$\alpha_{i,i} \ge \sum_{\substack{j=1\\ j\neq i}}^{n} |\alpha_{i,j}| \quad \textit{for all} \quad i \in N, \tag{13}$$

with strict inequality holding in (13) *for at least one i in N. Moreover,*

$$a_{i,j} = 0 \ \textit{ implies } \ \alpha_{i,j} = 0 \qquad \textit{(but not necessarily conversely)}. \tag{14}$$

Proof. For each $\epsilon > 0, A(\epsilon) := A + \epsilon I_n$ is a symmetric strictly ultrametric matrix. Hence from Theorem 1.1, $(A + \epsilon I_n)^{-1} := [\alpha_{i,j}(\epsilon)]$ in $\mathbb{R}^{n\times n}$ is a strictly diagonally dominant Stieltjes matrix, so that

$$\alpha_{i,j}(\epsilon) \leq 0 \text{ for all } i \neq j \text{ in } N \text{ and for each } \epsilon > 0, \tag{15}$$

$$\alpha_{i,j}(\epsilon) > \sum_{\substack{j=1\\ j\neq i}}^{n} |\alpha_{i,j}(\epsilon)| \text{ for all } i \in N \text{ and for each } \epsilon > 0, \text{ and} \tag{16}$$

$$a_{i,j}(\epsilon) = 0 \text{ if and only if } \alpha_{i,j}(\epsilon) = 0 \text{ for each } \epsilon > 0. \tag{17}$$

On letting $\epsilon \downarrow 0, A = A(0)$ is positive definite from Lemma 2.1, and its inverse, $A^{-1}(0) = [\alpha_{i,j}(0)]$ in $\mathbb{R}^{n\times n}$, is then well-defined. Again, letting $\epsilon \downarrow 0$ in (15) and (16) gives

$$\alpha_{i,j}(0) \leq 0 \text{ for all } i \neq j \text{ in } N, \text{ and} \tag{15$'$}$$

$$\alpha_{i,i}(0) \geq \sum_{\substack{j=1\\ j\neq i}}^{n} |\alpha_{i,j}(0)| \text{ for all } i \in N. \tag{16$'$}$$

That strict inequality must hold, for some i, in (16$'$) is clear, for otherwise, $A^{-1}(0)\xi_n = \mathbf{0}$, which contradicts the fact, from Lemma 2.1, that $A = A(0)$ and its inverse are both positive definite.

Finally, to establish (14), first note from (7*iii*) that no diagonal element $a_{i,i}$ of A can vanish, as this would force A to have a zero row and to be singular. Thus, the condition in (17) necessarily pertains only to off-diagonal entries of the matrices $A(\epsilon)$ and $(A(\epsilon))^{-1}$. But as $a_{i,j}(\epsilon) = a_{i,j}$ for all $i \neq j$ from (8), (17) can be expressed as

$$a_{i,j} = 0 \text{ if and only if } \alpha_{i,j}(\epsilon) = 0 \text{ for each } \epsilon > 0, \tag{17$'$}$$

and on letting $\epsilon \downarrow 0$, we can only deduce (cf. (14)) that

$$a_{i,j} = 0 \text{ implies } \alpha_{i,j}(0) = 0,$$

for it could be the case that $a_{i,j} > 0$ and $\alpha_{i,j}(\epsilon) > 0$ for each $\epsilon > 0$, while $\alpha_{i,j}(0) = 0$. □

3 An Example

Consider the matrix

$$A_1 := \begin{bmatrix} 3 & 3 & 2 \\ 3 & 4 & 2 \\ 2 & 2 & 3 \end{bmatrix}, \tag{18}$$

which can be verified to be a symmetric pre-ultrametric matrix in $\mathbb{R}^{3\times 3}$. Then, for each $\epsilon > 0$,

$$A_1(\epsilon) := A_1 + \epsilon I_3 = \begin{bmatrix} 3+\epsilon & 3 & 2 \\ 3 & 4+\epsilon & 2 \\ 2 & 2 & 3+\epsilon \end{bmatrix}, \tag{19}$$

whose associated rooted tree (cf. Theorem 2) can be verified to be

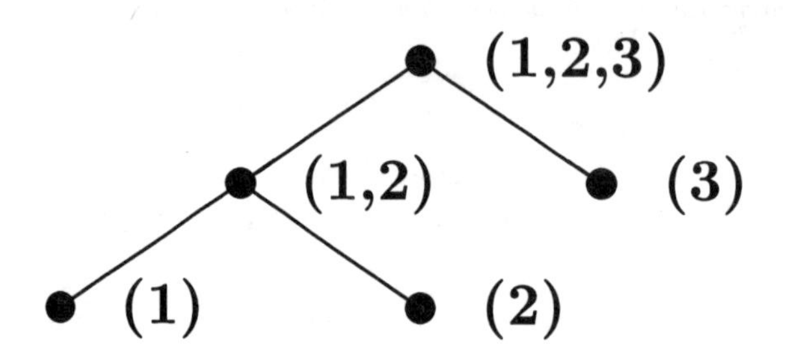

Then, the representation for $A_1(\epsilon)$ in (9) holds with the following definitions:

$$\begin{cases} \tau_1(\epsilon) := 2; \mathbf{u}_1 := (1,1,1)^T & \tau_2(\epsilon) := 1; \mathbf{u}_2 := (1,1,0)^T \quad \tau_3(\epsilon) := \epsilon; \mathbf{u}_3 := (1,0,0)^T \\ \tau_4(\epsilon) := 1+\epsilon; \mathbf{u}_4 := (0,1,0)^T & \tau_5(\epsilon) := 1+\epsilon; \mathbf{u}_5 := (0,0,1)^T. \end{cases}$$

In this case (cf. (10)),

$$\text{span } \{u_\ell : \tau_\ell(0) > 0\} = \text{ span } \{(1,1,1)^T; (1,1,0)^T; (0,1,0)^T; (0,0,1)^T\} = \mathbb{C}^3,$$

so that A_1 is then a symmetric ultrametric matrix. Now, the inverse of A_1, given by

$$A_1^{-1} = \begin{bmatrix} +1.6 & -1 & -0.4 \\ -1 & +1 & 0 \\ -0.4 & 0 & 0.6 \end{bmatrix}, \tag{20}$$

is such that A_1^{-1} has *some* zero off-diagonal entries, while A_1 has *no* zero off-diagonal entries. This shows, in particular, that the implication of (14), which holds vacuously for A_1, is true, but that the inverse implication in (14) does *not* hold. We also see from (20) that strict diagonal dominance holds in the first and third rows of A_1^{-1}, while only diagonal dominance holds in the second row of A_1^{-1}.

As a final comment, it was our original thought that the case of strict inequality in ($1iii$) could be weakened to the case of inequality in ($1iii$), while still preserving the main results of Theorem 1.1, if some irreducibility-like additional hypothesis were added. As it turned out, our additional hypothesis, that of (10), is what resulted. It is perhaps interesting to note that the assumption of (10) and the assumption of irreducibility of a matrix (cf. [3, p. 19]) both can be viewed as *global properties* of a matrix.

References

[1] S.Martínez, G. Michon, and J. San Martín, *Inverses of ultrametric matrices are of Stieltjes type.* SIAM J. Matrix Analysis and Appl. (to appear).

[2] R. Nabben and R.S. Varga, *A linear algebra proof that the inverse of a strictly ultrametric matrix is a strictly diagonally dominant Stieltjes matrix.* SIAM J. Matrix Analysis and Appl. (to appear).

[3] R.S. Varga, *Matrix Iterative Analysis.* Prentice-Hall, Inc., Englewood Cliffs, N.J., 1962.

Authors

1. Adam Bojanczyk, Cornell University, Department of Electrical Engineering, Ithaca, NY 14853-3801, USA.

2. Carlos F. Borges, Department of Mathematics, Naval Postgraduate School, Monterey, CA 93943, USA.

3. Daniela Calvetti, Department of Pure and Applied Mathematics, Stevens Institute of Technology, Hoboken, NJ 07030, USA.

4. Michael Eiermann, Institut für Praktische Mathematik, Universität Karlsruhe, D-7500 Karlsruhe 1, F.R.G.

5. K. Vince Fernando, NAG Ltd, Wilkinson House, Jordan Hill, Oxford OX2 8DR, UK.

6. Roland W. Freund, AT&T Bell Laboratories, Room 2C-420, 600 Mountain Road, Murray Hill, NJ 07974, USA.

7. William B. Gragg, Department of Mathematics, Naval Postgraduate School, Monterey, CA 93943, USA.

8. Martin Hanke, Institut für Praktische Mathematik, Universität Karlsruhe, Englerstrasse 2, W-7500, Karlsruhe, Germany.

9. Sangback Ma, Computer Science Department, 4-192 EE/CSci Building, 200 Union Street S.E., University of Minnesota, Minneapolis, MN 55455, USA.

10. Reinhard Nabben, Fakultät für Mathematik, Universität Bielefeld, D-4800 Bielefeld 1, FRG.

11. Noël M. Nachtigal, Research Institute for Advanced Computer Science, NASA Ames Research Center, Moffett Field, CA 94035, USA.

12. Beresford N. Parlett, Department of Mathematics, University of California, Berkeley, CA 94720, USA.

13. Johnny Petersen, IBM Bergen Scientific Centre, Thormøhlensgaten 55, N-5008 Bergen, Norway.

14. Robert Plemmons, Department of Mathematics and Computer Science, Wake Forest University, Box 7388, Winston-Salem, NC 27109, USA.

15. Lothar Reichel, Department of Mathematics and Computer Science, Kent State University, Kent, OH 44242, USA.

16. Youcef Saad, Computer Science Department, 4-192 EE/CSci Building, 200 Union Street S.E., University of Minnesota, Minneapolis, MN 55455, USA.

17. Richard S. Varga, Institute for Computational Mathematics, Kent State University, Kent, OH 44242, USA.

18. Paul Van Dooren, University of Illinois at Urbana-Champaign, Coordinated Science Lab, Urbana, IL 61801, USA.